44th Annual Meeting of the Plant Growth Regulation Society of America 2017

Anchorage, Alaska, USA
6 – 10 August 2017

ISBN: 978-1-5108-5712-4

Printed from e-media with permission by:

Curran Associates, Inc.
57 Morehouse Lane
Red Hook, NY 12571

Some format issues inherent in the e-media version may also appear in this print version.

Printed by Curran Associates, Inc. (2018)

Plant Growth and Regulation Society of America
1018 Duke Street
Alexandria, VA 22314
USA

Phone: 703-836-4606 x116
Fax: 703-836-4607

cslone@ashs.org

Additional copies of this publication are available from:

Curran Associates, Inc.
57 Morehouse Lane
Red Hook, NY 12571 USA
Phone: 845-758-0400
Fax: 845-758-2633
Email: curran@proceedings.com
Web: www.proceedings.com

Proceedings

of the

44th Annual Meeting

of the

Plant Growth Regulation Society of America

Hilton Downtown Anchorage

Anchorage, Alaska

August 6-10, 2017

Proceedings of the 44th Annual Meeting of the Plant Growth Regulation Society of America

MEETING SUMMARY

44th Annual Meeting of the Plant Growth Regulation Society of America

Jozsef Racsko
Meeting Convener & 1st Vice President, PGRSA

The 44th Annual Meeting of the Plant Growth Regulation Society of America was held in the heart of historic downtown Anchorage, Alaska. The Steering Committee has chosen this site for the meeting because of the unique location and the beauty of the environment. This Conference served as a great learning and networking opportunity for all participants and had an outstanding group of invited speakers to set the stage for a dynamic program with presentations from academic, government and private company research and outreach programs

Prior to the official kick-off of the meeting, participants, who chose to do so, attended the pre-conference tour, experiencing something uniquely Alaska – William's Reindeer Farm! While attendees enjoyed the local foods and drinks at the Welcome Reception, local extension agent and mountaineer, Stephen Brown, shared his experiences climbing Alaska's famed peak – and North America's tallest – Mount Denali. Stephen has not only climbed Denali twice, but has summited many of the world's tallest mountains, including Kilimanjaro.

The meeting featured a wonderful lineup of keynote speakers and contributed presentations, including Miriam Karlsson of the University of Alaska, Emily Merewitz from Michigan State University, Steve West of Research Designed 4 Agriculture, Alok Adholeya of the TATA Energy Research Institute, Maria Herrero from Valent BioSciences Corporation and David Beudreau of DC Legislative and Regulatory Services. This year's Historic Perspectives talk was delivered by James Barrett of the University of Florida on the evolution of the use of PGRs in ornamentals. In addition, our Young Scientist of the Year award winner, Caroline Gutjahr, from the Ludwig Maximilian University, Germany presented her research on phytohormone signaling in arbuscular mycorrhiza development. This year's meeting attracted eight students to participate in the poster and oral presentation competitions, and four student travel awards were given this year.

The program committee members for the meeting included Casey Barickman, Gurreet Brar, Chris Gunter, Holly Little, Sonja Maki, Nicole Waterland, Tim Spann, Marcus Jones, Jozsef Racsko, and Michael Rethwisch. All of these members are to be commended for their hard work and creativity in putting together a fantastic program.

On behalf of the PGRSA Steering Committee, I cordially invite everyone interested in plant sciences to attend the 45th Annual Conference of the PGRSA to be held in San Juan, Puerto Rico, June 10 - 14, 2018. We look forward to seeing you there!

HONORARY LIFETIME MEMBERS

Anson Cook
Aldo J. Crovetti
Eric A. Curry
Horace G. Cutler
Richard Dunand
C. Dave Fritz
Masayuki Katsumi (JSCRP)
Wayne Mackay
Wilhelm Rademacher
Edward F. Sullivan
Thomas Tworkoski

Acknowledgements

The Proceedings has been published yearly since the fourth annual meeting in 1977 at no charge to the author(s). Papers in this edition were received electronically and, although we have made minor changes in formatting, every effort has been made to preserve the original submissions, the content of which remains the sole responsibility of the respective author(s).

The Proceedings were compiled by:

Dr. Jozsef Racsko, 1st Vice President, Meeting Organizer
Mycorrhizal Applications, LLC
710 NW E Street
Grants Pass, OR 97526
Office 541-956-4116
Cell 224-456-1366
jozsef.Racsko@mycorrhizae.com

The Proceedings are available for download at www.PGRSA.org. The most recent two years of the Proceedings are available to PGRSA members only. Older issues are available to everyone free of charge.

Hard copies of the Proceedings are available for a fee from Curran Associates, Inc. print on demand services. To order a hard copy visit http://www.proceedings.com/.

A SPECIAL THANK YOU TO THE FOLLOWING COMPANIES
FOR FINANCIAL SUPPORT OF THE MEETING

Platinum Sponsor

VALENT BIOSCIENCES CORPORATION

Mycorrhizal Applications

Gold Sponsors

fine

Silver Sponsors

WINFIELD UNITED

AMVAC
Amvac Chemical Corporation

Bronze Sponsors

ohp
Partners with solutions

REPAR

2016 – 2017 PGRSA Steering Committee

Dr. Chris Gunter, President
Horticultural Science, North Carolina State University
230 Kilgore Hall, Box 7609
Raleigh, NC 27619-7609
Office 919-513-2807
Cell 919-368-2724
Fax 919-515-2505
chris_gunter@ncsu.edu

Dr. Holly Little, Past President
Acadian Seaplants Ltd.
2443 Fair Oaks Blvd., #67
Sacramento, CA 95825
Office 530-219-5415
hlittle@acadian.ca

Dr. Jozsef Racsko, 1st Vice President
Valent Biosciences Corp.
870 Technology Way
Libertyville, IL 60048
Office 847-968-4720
Cell 224-456-1366
jracs@valent.com

Dr. Gurreet Brar, 2nd Vice President
Assistant Professor
Department of Plant Sciences
California State University
2415 E. San Ramon, Fresno, CA 93740-8033
Office 559-278-4119
Cell: 239-290-5666
gurreetbrar@csufresno.edu

Mr. Michael Rethwisch, Secretary
University of Nebraska – Lincoln
451 N. 5th Street
David City, NE 68632
Office 402-367-7410
mrethwisch2@unl.edu

Dr. Marcus Jones, MAL-1
WinField Solutions, LLC
2777 Prairie Drive
River Falls, WI 54022
Office 515-520-7815
majones@landolakes.com

Dr. Nicole Waterland, MAL-2
Division of Plant and Soil Sciences
West Virginia University
Morgantown, WV 26506-6108
(304) 293-2969
Nicole.Waterland@mail.wvu.edu

Dr. Casey Barickman, MAL-3
North Mississippi Research and Extension Center
Mississippi State University
PO Box 1690
Verona, MS 38879
Office 662-566-8001
tbarickm@ext.msstate.edu

Dr. Tim Spann, Business Manager
California Avocado Commission
12 Mauchly, Suite L
Irvine, CA 92618
Office 949-341-1955
Cell 863-258-5950
tim@pgrsa.org

Dr. Sonja L. Maki, Executive Director
Biology Department, Carleton College
One North College Street
Northfield, MN 55057
Cell 715-690-4334
Fax 507-222-5757
smaki@carleton.edu

Ms. Cindy Slone, Executive Secretary
PGRSA/American Society for Horticultural Science
1018 Duke Street
Alexandria, VA 22314
Office 703-836-4606
Fax 703-836-4607
cslone@ashs.org

TABLE OF CONTENTS

LOCALLY GROWN: FOOD SECURITY AND HORTICULTURE IN ALASKA

Meriam Karlsson
University of Alaska Fairbanks, 303 O'Neill Bldg., Fairbanks, AK

Short growing seasons, limited infrastructure, and high energy costs can be problematic for producing crops in Alaska. Research to extend production throughout the year using efficient production systems and technology is necessary to improve local fresh food availability and overall food security. Although horticulture production can be challenging, Alaska also offers unique opportunities for research into crop growth, development and management in a northern environment. For instance, effects of climate change have already been documented and the naturally large variation in day lengths and temperatures over the year allows for environmental stress and interaction studies. Information related to food availability, marketing and distribution in market- as well as subsistence-based systems has applications for rural food insecurities throughout the United States and Canada.

Alaska is the largest state in the United States with a land area of approximately 375 million acres. To put the size of the state in perspective, flying from Juneau in Southeast Alaska to Barrow on the Arctic Ocean is about 1200 air miles or the corresponding distance from Orlando to New York. A drive from Homer to the Prudhoe Bay airport of Dead Horse through Anchorage and Fairbanks via Parks, Elliot and Dalton highways, is more than 1,000 miles and likely to take more than 24 hours. A matching trip in the contiguous United States would be to drive from the Canadian border to Tijuana, Mexico through Washington, Oregon and California.

Although the land area is large, the state's population is among the smallest. Only Wyoming and Vermont have a more limited population than Alaska at the 2016 estimate of 742,000. The large geographical area with a small disperse population, variable climatic and weather conditions along with limited infrastructure, road systems and transportation alternatives, result in potential difficulties distributing food and other goods throughout the state and to the many often small and remote communities.

USDA defines food security in the United States as "access by all people at all times to enough food for an active, healthy life" (https://www.ers.usda.gov/topics/food-nutrition-assistance/food-security-in-the-us/). In addition, USDA considers four levels of food security as high, marginal, low and very low. Regions of very low food security will have disruptions in eating patterns, reduced food intake and hunger. Considering food security on a community rather than individual basis is often more accurate for rural areas where people tend to be more dependent on each other, social networks and group activities for daily life and survival. Hamm and Bellows (2003) defined community food security as "a community where all residents obtain a safe, culturally acceptable, nutritionally adequate diet that maximizes community self-reliance and social justice." Safety and cultural considerations are often significant components of food security in rural Alaska villages with long cultural traditions of subsistence hunting, fishing, gathering and cultivation.

State legislatures are aware that Alaska is vulnerable in respect to food security. To better understand the specific issues and dispel misconceptions, the State through the Alaska Department of Health and Social Services commissioned Crossroads Resource Center in Minneapolis, to document current conditions and provide recommendations for future actions. The outcome of this study is available in the report "Building Food Security in Alaska" by Meter and Phillips-Goldenberg (2014).

During information gathering and discussions with agency and university employees, community leaders and the general public, most Alaskans defined and viewed food security as "a secure food supply that is protected from disruptions caused by for instance, weather, flooding, strikes, political conflicts, unreliable supply lines, transportation issues" (Meter and Phillips-Goldenberg, 2014). Since at least periodically local sources of food are limited, many Alaskans realize the high dependence on outside production, processing and transportation for food availability. Even short incidences of transportation disruptions such as the 2012 washout of the Alaska Highway in Canada caused immediate shortage of certain food items such as milk, eggs and vegetables.

USDA ranks Alaska last compared to other states in agricultural production. According to the 2012 Census of Agriculture, the agricultural farm gate value was $31 million. Production at many small and often remotely located farms may not be counted and included in the official value of the Alaska agricultural production. Still, the production is small and even Rhode Island with a land area of 1,212 square miles compared to Alaska's 663,000 square miles, has a significantly larger agricultural industry at $58 million than Alaska. It may be of interest to note that the sector with the largest value in Alaska is greenhouse and nursery production. Although greenhouse and nursery crops have been a significant portion of Alaska agriculture for many years, detailed information on crops, area and type of production is usually not available.

Additional facts about agriculture and food security documented in the report on building food security in Alaska (Meter and Phillips-Goldenberg, 2014) suggest that 95 percent of the purchased food comes from outside the state. Alaskans buy food each year for an estimated $2 billion (possibly $2.5 billion). Subsequently, Alaskans spend at least $1.9 billion each year on food from outside the state. Keeping more of these dollars in community economies would strategically build food security and local agricultural capacities throughout the State.

Meter and Phillips-Goldenberg (2014) provide several recommendations for how to improve food security in Alaska. The first recommended action is to expand agricultural production and gardening along with building agricultural capacities. This is a logical suggestion considering the current small production volume for increasing self-reliance and better prepare for potential interruptions in food availability. Several efforts have been made over the years to establish and expand production agriculture. Production efforts for local markets is now the emphasis rather than targeting export, as was the case in some earlier less successful projects. The current local production activities in several Alaska regions are showing encouraging progress.

The USDA NRCS Seasonal High Tunnel Initiative (https://www.nrcs.usda.gov/wps/portal/nrcs/detailfull/national/programs/?cid=stelprdb1046250) has been and continuous to be successful in Alaska. The goals of the Seasonal High Tunnel Initiative are applicable to local growing and marketing conditions and Alaska was one of the pilot states when the high tunnel program was introduced in 2010. Goals for the initiative include extending the growing season, improve plant and soil quality, reduce nutrient and pesticide use, improve air quality through reduced transportation input, and reduce the energy use by providing consumers with a local source of fresh produce.

Information from USDA/NRCS indicates that Alaska is one of the top states taking advantage of the high tunnel initiative along with Missouri, Minnesota, Wisconsin and Michigan. In 2016, close to 700 Alaska high tunnels had been established through this program. Some regions in Alaska have especially been successful in utilizing the high tunnel program such as the Kenai Peninsula. The community of Homer for instance with a population of 5,000, has more than 200 tunnels in crop production. High tunnels are also used in many other areas such as the Kodiak Island, Interior Alaska, south-central, southeast, and western Alaska around Kotzebue. The main purpose of using high tunnels is to extend and enhance the summer growing season. During the winter-half of the year, the plastic cover is usually removed to avoid snow and wind damage. Melting snow also allows for leaching and control of soluble salt levels in the high tunnel soil.

Along with the interest and use of high tunnels, many people in Alaska are actively involved with gardening. A survey in 2013 on the number of community gardens in relation to state population, puts Alaska only behind North Dakota, Nebraska and Iowa. Considering the short growing season, many small communities and challenging growing conditions, the high community garden ranking can be considered significant and to some extent surprising. There are also records to suggest that every third Alaskan regularly eats something that was grown in their own garden (Alaska Department of Health and Social Services, 2013).

Focus on consumer attention and loyalty to Alaska grown food is another action recommended by Meter and Phillips-Goldenberg (2014). The Alaska Grown certification is a promotional marketing program (http://dnr.alaska.gov/ag/ag_AKGrown.htm). Initiated in the 1980, the Alaska Grown program is administered by the Division of Agriculture in the Alaska Department of Natural Resources. The program operates with a limited budget in managing approvals and compliance of logo use. The Alaska Grown label indicates that the product is locally grown in Alaska. Assumptions of organic production methods are not uncommon among the public, but production techniques are not considered or evaluated in the Alaska Grown certification process. The limited marketing program supports special promotions such as the 5 for 5 Alaska grown challenge (spend $5 weekly on Alaska Grown products for five months), professional cooking demonstrations at Farmers Markets, or specific advertisement campaigns such as the Fresher by Far campaign. In general, though, advertising efforts are limited as the demand for Alaska grown products often outpace local availability.

In addition to the Alaska Grown program, local community marketing labels are starting to appear. The Golden Heart Grown program for instance, was launched this year in Interior Alaska for the Fairbanks area (https://investfairbanks.com/sites/default/files/documents/GHG%20App%20and%20Reqs1.pdf). These efforts continue to also support the Alaska Grown marketing program while developing and creating closer local community connections.

Closely connected to the Alaska Grown efforts are the Farm to School program (http://dnr.alaska.gov/ag/ag_FTS.htm). Also housed in the Division of Agriculture, the Farm to School program's main purpose is to increase the amount of Alaska grown products served in schools. Despite scheduling discrepancies between seasonal production and school sessions, the program has successfully increased locally grown products not just in schools but also in other state-run institutions. In communities where fishing is a major source of income, the companion program Fish to School is often more appropriate for increasing local food consumption and awareness. In addition to the incorporation of local foods, the Farm- and Fish to School programs provide educational resources for teachers on food choices, nutritional values and availability.

To increase the understanding of food options, choices and production, many schools in Alaska have established greenhouses. In several communities, schools are heated with biomass as this is a readily available local resource. Greenhouses appear to be a good addition and opportunity when the capacity of installed boilers often exceed the need of the main school buildings. Many of these schools incorporate greenhouse lessons in support of the overall curriculum. Several of the greenhouses also produce crops for the school cafeteria or local restaurants. "The Biomass-Heated Greenhouses Handbook, a handbook for Alaskan schools and community organizations" is sponsored by the Alaska Energy Authority and was recently released. It is available on-line at http://www.akenergyauthority.org/Programs/AEEE/Biomass.

Like national trends, direct marketing such as farmers markets, community supported agriculture (CSA), subscription farming and farm stands, has become a major outlet for producers in the state. Direct sales increased 32 percent over five years in Alaska and now bring in one of every five dollars earned by food producing farmers. The number of farmers markets have increased to more than 50 with a revitalized Alaska Farmers Market Association (https://alaskafarmersmarkets.org). Various farms offer subscriptions through CSA programs. Many of these farms are small providing a limited number of shares and currently there is more demand than availability for CSA. The reasons for subscribing to a CSA program are likely similar to national trends and include growing Alaska's local economies, building community relationships, knowing where the food comes from, Alaska food tastes better, and regular availability of locally grown to build healthy eating habits.

The most successful local food production endeavors tend to be on a small scale. These efforts would be stronger if Alaska created food processing and manufacturing facilities for in-state markets. State and privately-run facilities have had variable success over the years and local production is often too small to justify specialized food processing. This creates a dilemma, as processing facilities are necessary to increase local production while there isn't enough local product to justify the establishment of suitable infrastructure.

Other recommended actions for improving food security include adopting state policies to support local food production and strengthen internal food distribution networks with statewide transparency and coordination (Meter and Phillips-Goldenberg, 2014). The Alaska Food Policy Council (AFPC, https://www.akfoodpolicycouncil.org) is an independent non-profit organization.

Groups and people interested in food security and other issues related to food production, consumption and availability founded AFPC in 2010. It now has members and participants from more than 200 agencies along with private citizens. In bringing various groups and interested parties together to talk about food related issues, drafting policies and suggesting state governmental actions, the AFPC has been significantly more successful than any committee or task force created by official Alaska state offices. Several active regional food system organizations have also developed over the last few years such as Sustainable Juneau, Kenai Local Food Connection and Sitka Local Foods Network.

The Alaska Food Coalition (http://alaskafood.org/) is a food security and hunger program that unites 125 non-profit and state agencies. The goal is to strengthen and coordinate public and private food nutrition assistance programs. An additional goal is to educate policy makers about the needs for providing food in and to Alaska. Food Bank of Alaska (www.foodbankofalaska.org) and various local food banks throughout the state actively work to aid families, individuals and areas with limited food resources.

Subsistence is a key component of food security in Alaska and becomes more important the further away a community is from the road system. Meter and Phillips-Goldenberg (2014) recognize this importance and recommend continued promotion and fostering of subsistence harvesting and related skills. Many agencies, organizations and researchers are actively working on various aspects related to subsistence. However, subsistence is a multifaceted and complicated area with variable dimensions of acquiring, preparing, sharing, storing, human nutrition, traditional ways of living, cultural impacts, social importance, community development and economic growth.

Personal use hunting, fishing and gathering are a significant food source not just for native and rural communities, but also urban populations. An estimated 65 percent of the Alaska population, harvest wild foods at least once every year. The protection and management of ecosystems are essential to sustain and enhance subsistence resources. Grounds used for hunting, fishing and gathering are often a combination of federal, state and native owned lands. Conflicts related to the management and use are not uncommon between federal and state agencies. Local and native knowledge and interests have many times in the past been ignored. Even though, people using and living on the land can be expected to have the most recent understanding of current trends and issues. Several efforts are under way to co-manage resources for subsistence and personal use harvesting that incorporates local knowledge along with assistance and input from state and federal agencies.

The Alaska Native Tribal Health Consortium (https://anthc.org) is a non-profit health organization working with rural populations. ANTHC has established and run several health centers around the state. There is often a need for specialized approaches, services and assistance, as health issues tend to be different in rural areas compared to the general population.

Access to diets that are nutritionally balanced, culturally acceptable and affordable is a significant component in determining overall health and human wellbeing. The "Store Outside Your Door" is a program run by ANTHC with the mission to document current knowledge, develop educational resources and promote the use of traditional ways and foods. The approach to subsistence is changing as modern equipment, technology and knowledge are becoming available. For instance, a common concern now among communities dependent on subsistence foods is the price of fuel as power boats, snow machines and other vehicles are used more extensively. Changes in climate and weather patterns are affecting the landscape, vegetation types, migration paths, ice formation and road access. The techniques and approaches used for hunting, fishing, gathering and cultivation need to continuously adapt to these changing ecosystems.

Although it may be challenging to keep Alaska food secure, there are also distinctly unique opportunities for local production, innovations and food sources. The field season is short but extended day lengths and moderate temperatures support high quality plant growth. For instance, the long days allow for extended daily periods of photosynthesis without stomata closure in the middle of the day due to high temperatures.

The isolation from large agricultural production areas and long cold winters tend to reduce the pest and disease pressure. Common pests such as aphids, thrips and spider mites do appear along with diseases such as gray mold and powdery mildew. Still compared to intense agricultural regions, pests tend to be more limited and manageable. In addition, some pests do not usually appear in Alaska such as the Colorado potato beetle or late blight.

The Alaska summer growing conditions with extended daily periods of photosynthesis often result in large sized vegetables and flowers. The more intense flower and fruit colors can be contributed to the moderate temperatures and cooling during the night. Cool fall nights combined with optimal daily growing conditions often also result in vegetables with high sugar content and enhanced flavor. Carrots for instance, tend to be sweeter than those grown in other locations. High nutrient levels have been recorded with exceptionally high antioxidant contents in Alaska blueberries.

Food security during the winter half of the year can be problematic. In Fairbanks (64.81°N latitude), the time between sunrise and sunset during the shortest day in December is 3 hours and

40 minutes. Temperatures dropping to -40°F are not unusual. During these conditions, crops can only be produced in controlled environments such as greenhouses. The main limiting factor for this type of production is energy costs for heating and lighting. Although the Alaska Energy Authority (http://www.akenergyauthority.org) has a mission of reducing the cost of energy in Alaska, energy costs continue to be high.

The most sensible approach during the winter half of the year may be to produce crops in well insulated controlled environments with advanced technology such as light emitting diodes. However, six months out of the year, Alaska has more than twelve hours of natural day lengths. Growing crops in facilities that do not take advantage of these prolonged daily light conditions may make less sense. Utilizing the summer field environment to the fullest combined with visions of processing, preserving and storing the produced food for the remainder of the year, may be one of several needed methods to make Alaska more food secure.

REFERENCES

Alaska Department of Health and Social Services. 2013. Alaska obesity facts: local foods. Available at: http://dhss.alaska.gov/dph/Chronic/Documents/Obesity/pubs/factsheet_LocalFoods.pdf

Hamm, M.W. and A.C. Bellows, 2003. Community food security and nutrition educators. J. Nutritious Education and Behavior Engagement 35:37-43.

Meter, K. and M. Phillips-Goldenberg. 2014. Building Food Security in Alaska. Crossroads Resource Center, Minneapolis. Available at: https://static1.squarespace.com/static/584221c6725e25d0d2a19363/t/59a04ad6f9a61e091b836459/1503677150479/Building+Food+Security+in+AK_Ken+Meter_July+2014_web+version.pdf

METABOLIC AND MOLECULAR FACTORS FOR HORMONAL REGULATION OF HEAT-INDUCED LEAF SENESCENCE

Bingru Huang
Department of Plant Biology, Rutgers University, New Brunswick, NJ 08901

Leaf senescence is a typical characteristic of heat stress, which attributes to the reduction of photosynthesis and plant growth. It is well known that hormones play critical roles in regulating lead senescence. Understanding metabolic mechanisms and molecular factors for hormonal regulation of heat-stress induced leaf senescence is of greatly important for manipulating hormone metabolism to control stress-induced leaf senescence. We conducted studies with exogenous application of different hormones and genetically modified plants of perennial grass species to identify major hormones controlling heat-induced leaf senescence. Our studies strongly suggest that cytokinins play the key role in suppressing heat-induced leaf senescence in perennial grass species. The suppression of heat-induced leaf senescence is largely associated with the down-regulation of genes and deactivation of enzymes in the metabolic pathways of chlorophyll degradation. Several cytokinin- responsive transcriptional factors and down-stream genes are also been identified.

EFFICACY OF 1-MCP IN GROWTH MODULATION OF TOMATO AND PEPPER TRANSPLANTS EXPOSED TO HEAT STRESS

Daniel I. Leskovar, Yahia Othman and Desire Djidonou
Texas A&M AgriLife Research, Texas A&M University, Uvalde, Texas, USA

Vegetable growers require high quality transplants able to withstand abiotic stresses to ensure a high field survival and rapid growth after transplanting. The application of the ethylene inhibitor 1-methylcyclopropene (1-MCP) to modulate growth of vegetable transplants under stress temperatures is unknown. The objective of this study was to assess the influence of 1-MCP on growth and quality of tomato and pepper transplants exposed to stress temperatures. Eight tomato cultivars representing round, plum, cherry and home-grown types, and six pepper cultivars representing bell, jalapeno and long green chili types were grown in greenhouse nursery conditions for six to seven weeks and then treated with 1-MCP (50 $mg \cdot L^{-1}$) using a boom fertigation system. After treatments, both 1-MCP and untreated plants were transferred to 0.5 L pots and placed in environments differing in day/night temperature regimes. For tomato, we used 30/20 °C and 34/24 °C constant, and 34/24 °C ramping temperatures; and for pepper, the temperature regimes were 24/18 °C and 32/23 °C constant, and 35/25 °C ramping. Thereafter, plant morphology and physiology variables were collected during the study period. In tomato, 1-MCP application significantly increased leaf area, stem diameter and plant height of cv. Florida 47 (round type) across all temperature regimes. Also, cv. TAMU-Hot (home-grown type) which was grown only in the 34/24 °C (ramping) also responded positively to 1-MCP. In pepper, 1-MCP significantly enhanced the total biomass of cv. LARK 1845 (jalapeno) while significantly reduced the biomass of cv. PS0 994 18 19 (bell) under the 24/18 °C regime. Overall, 1-MCP had both, promotive and depressing effects in tomato seedlings at normal and heat stress temperatures, while the effects of 1-MCP on pepper types were minimal.

NEW APPROACHES AND USES OF PLANT GROWTH REGULATORS IN TURFGRASSES

Emily Merewitz
Michigan State University, East Lansing, MI 48824

Plant growth regulators (PGRs) are widely used in turfgrass management. Identification of new compounds of utility to the turfgrass industry is needed due to market loss of available chemistries, changing environmental factors and diseases, and high-performance expectations of turfgrass species. Polyamines (PAs) have been shown to have PGR effects and play a role in stress tolerance in some crop species. Prior to our research, it was not known how PAs may affect important turfgrass species. Therefore, a series of physiological and molecular experiments to understand the mechanism associated with PAs effect on growth and abiotic stress tolerance of creeping bentgrass (*Agrostis stolonifera*) will be discussed. In addition to investigation of new compounds, a major research objective is to evaluate new uses for PGRs or fungicides, such as abiotic or biotic stress tolerance, already in use in the turfgrass industry. For instance, we have investigated whether Civitas, mefluidide, propiconazole, and trinexapac-ethyl (TE) may promote cold acclimation and tolerance of ice encasement. Annual bluegrass (*Poa annua*), which is susceptible to winterkill damage, treated with mefluidide, propiconazole, and Civitas had a greater percentage of polyunsaturated FAs, with linoleic acid being increased by approximately 50% due to these chemical treatments compared with TE and untreated samples. Results from this study suggest that applications of Civitas, mefluidide, or propiconazole in the fall season may improve ABG survival of severe winter conditions. This survival may be associated with a shift in the FA profile towards more polyunsaturated FAs in crown membrane tissue.

TRANSCRIPTOME ANALYSIS OF PETUNIA GROWN UNDER WATER DEFICIT

Suejin Park, Youyoun Moon, and Nicole L. Waterland*

Division of Plant and Soil Sciences, West Virginia University, Morgantown, WV 26506

Water deficit is a major constraint to plant growth and development that could result in severe crop loss. To combat damages from water deficit stress, development of crops with an increased tolerance to water stress is highly desirable. Physiological and molecular experiments have been performed in various crops and a significant understanding of genes that are involved in the water deficit stress mechanism has been achieved. Despite of a recent advancement of genomic research techniques, genome wide investigation of water stress response mechanism in horticultural crops is still scarce. In our study, RNA sequencing was performed to generate transcriptome data of Petunia × hybrida 'Mitchell Diploid' under water deficit stress. The objectives of this research were to 1) analyze transcriptome data to provide a global gene expression profile, and 2) identify candidate genes for functional analyses. Nine-week-old petunias were irrigated daily or placed under water stress by withholding water. Leaf tissue samples were collected 1, 3, and 5 days after water was withheld. Nearly 270 million reads were sequenced by Illumina Hiseq 1500, and De novo assembly using Trinity software generated 76,601 contigs. Among all contigs, 154, 3611, and 980 contigs were upregulated and 41, 2806, and 253 contigs were downregulated on day 1, 3, and 5, respectively. Gene Ontology (GO) annotation was carried out by Blast2GO. Among the differentially expressed genes, genes encoding regulatory proteins such as transcription factors, kinases and phosphatases were identified. Six transcription factors that were upregulated at the early stage of water deficit stress were chosen for functional analyses.

GROWTH CONTROL OF TWO FLORICULTURE SPECIES USING PACLOBUTRAZOL LINER SOAKS

Josh B. Henry, Brian E. Whipker, and Ingram McCall
Department of Horticultural Science, North Carolina State University, Raleigh, NC 27606 U.S.A.

Applying plant growth regulators (PGRs) as a pre-transplant liner soak offers floriculture growers an effective alternative method of growth control in their crops. Dahlia (*Dahlia* 'Dahlianova Hypnotica Purple Bicolor') and petunia (*Petunia atkinsiana* 'Crazytunia Razzmatazz') were considered two vigorous crops and rooted liners were soaked in paclobutrazol (Piccolo 10 XC; Fine Americas) concentrations of 4, 8, 16, 32, and 64 mg·L^{-1}. These plants were soaked for two minutes and transplanted after 24 hours into 800-mL (12.7-cm diameter) pots. Plants were grown out and biweekly measurements were used to determine optimal paclobutrazol concentrations for each species. Pre-transplant liner soaks were an effective method of growth control for petunia, but only moderately so for dahlia. Optimal paclobutrazol concentrations for petunia were 4 – 8 mg·L^{-1} while dahlias required >32 mg·L^{-1}.

INTRODUCTION

Pre-transplant liner soaks of plant growth regulators (PGRs) are an alternative method growers can use to control growth in vigorous floriculture crops (Blanchard and Runkle, 2007). Previous studies investigating PGR liner soaks reported effective growth control in several species using paclobutrazol, uniconazole (Blanchard and Runkle, 2007), and flurprimidol (Whipker et al., 2003). Several factors are reported to impact the efficacy and plant growth response to liner soaks including the soak duration, the depth of the solution, and the time between treatment and transplant (Schnelle et al., 2005). For these reasons, it is imperative that PGR liner soak recommendations be made given the specific treatment and timing related factors. This study was done to determine optimal paclobutrazol concentrations for liner soaks of dahlia (*Dahlia*) and petunia (*Petunia atkinsiana*).

MATERIALS AND METHODS

'Dahlianova Hypnotica Purple Bicolor' dahlia and 'Crazytunia Razzmatazz' petunia cuttings (Dümmen Orange, Columbus, OH) were stuck in 128-cell plug flats (2.7l x 2.7w x 3.8d cm) filled with Sunshine® Mix #1 (Sun Gro Horticulture, Agawam, MA) on 5 April 2017. The cuttings were rooted under mist in a glass glazed greenhouse in Raleigh, NC (35° N Latitude). On 4 May, rooted liners were treated with a paclobutrazol (Piccolo 10XC; Fine Americas, Walnut Creek, CA) soak of 0, 4, 8, 16, 32, or 64 mg·L-1. Liners were soaked in the paclobutrazol solution for two minutes with 75% of the root system submerged and were then placed on a bench. Soaked liners were transplanted 24 hours after treatment into 12.7-cm diameter pots (Poppelmann Plastics USA LLC, Claremont, NC) filled with Sunshine® Mix #1. Each species was randomized and thoroughly watered with clear water. Plants were then placed on automatic irrigation and fertilized with a 13N – 0.86P–10.8K fertilizer mixed at 150 $mg{\cdot}L^{-1}$N. Biweekly measurements were taken for plant height by measuring from the substrate to the highest point on the plant. Diameter was measured by averaging the widest point of the plant and the axis perpendicular to it. Final data were collected six weeks after transplant.

Data were analyzed with SAS (version 9.4) using PROC MIXED to determine differences in growth over time, and least squared means were separated using the Tukey adjustment. Data were then subjected to PROC REG and PROC NLIN to determine the best fit linear, quadratic, or quadratic plateau model based on r2 and the corrected Akaike Information Criterion (AICc). Regression models were then used to find the vertex for quadratic models or X0 value where the model plateaus for quadratic plateau models. These values indicate at what paclobutrazol concentration each growth parameter was minimized.

RESULTS AND DISCUSSION

Pre-transplant paclobutrazol liner soaks had a moderate effect on both height and diameter of dahlia plants over the course of six weeks (Fig. 1). Two weeks after treatment, there were no significant differences in height, and only plants treated with 8, 32, and 64 mg·L-1 paclobutrazol had smaller diameters than control plants (Table 1). Four weeks after transplant, dahlias treated with 8 – 64 mg·L-1 paclobutrazol were smaller than control plants in height, while only those treated with 32 and 64 mg·L-1 paclobutrazol were smaller in diameter (Table 1). At termination, plants treated with 32 and 64 $mg{\cdot}L^{-1}$ paclobutrazol were smaller in height, and all diameters were similar (Table 1).

Regression analysis at each week indicated that dahlia plants exhibited a negative linear or quadratic response to increasing paclobutrazol concentrations in terms of both height and diameter (Fig. 2A). After two and six weeks, height and diameter exhibited a linear response to increasing paclobutrazol concentrations (Fig. 2A). After four weeks, height and diameter exhibited a quadratic growth response (Fig. 2A and C). All the models indicated that growth was minimized with paclobutrazol concentrations ≥53 mg·L^{-1} (Table 2). This demonstrates that although pre-transplant liner soaks provided some growth control, 'Dahlianova Hypnotica Purple Bicolor' is highly insensitive to paclobutrazol when applied as a liner soak.

Previous studies investigating paclobutrazol liner soaks determined optimal paclobutrazol concentrations of 4 – 12 mg·L^{-1} among several species (Blanchard and Runkle, 2007). Similar results were reported by Whipker et al. (2003), except for angelonia which required ≥16 mg·L^{-1} paclobutrazol. In contrast, Latimer and Groover (2005) recommended paclobutrazol liner soak concentrations of just 2–4 mg·L^{-1}. These concentrations are far lower than what was necessary to provide significant growth control for dahlia. Higher paclobutrazol concentrations may be required for growth control in dahlia, or another PGR may have greater efficacy.

Petunias were more sensitive to pre-transplant paclobutrazol liner soaks than dahlias. There were no significant differences in petunia height or diameter after two weeks (Table 1). After four weeks, petunias treated with 16 – 64 mg·L^{-1} paclobutrazol were shorter than control plants, and plants treated with 32 or 64 mg·L^{-1} paclobutrazol were smaller in diameter (Table 1). After six weeks, differences among paclobutrazol concentrations were visually apparent (Fig. 3). Height of plants treated with 16–64 mg·L^{-1} paclobutrazol remained shorter than control plants (Table 1). In fact, plants grown for six weeks after a 16 or 64 mg·L^{-1} paclobutrazol liner soak were similar in height to control plants two weeks after treatment (Table 1). All petunias that received a paclobutrazol liner soak were smaller in diameter than control plants, indicating significant growth control (Table 1).

Regression of petunia growth was fit best with either quadratic or quadratic plateau models at each week (Fig. 2B and D). At two weeks after treatment, a quadratic plateau model provided the best fit for petunia height and diameter (Table 2). These plateaus indicated that petunias required 4.7 and 6.9 mg·L^{-1} paclobutrazol to minimize height and diameter, respectively (Table 2). After four weeks, a quadratic model provided the best fit, and the saturation point for each model increased to 49.7 and 61.2 mg·L^{-1} paclobutrazol to minimize height and diameter, respectively (Table 2). After six weeks, a quadratic model was selected for height while a quadratic plateau model was selected for diameter (Table 2). These models indicated that height was minimized with 45.8 mg·L^{-1} paclobutrazol, while diameter was minimized with 16.4 mg·L^{-1} paclobutrazol (Table 2). The selected models and their corresponding saturation points illustrate how the PGR effect lessened over time.

Trends illustrated by these two species enable growers to make better predictions for how liner soaks of different paclobutrazol concentrations will affect their crops over time. This study demonstrated that dahlia required ≥32 mg·L^{-1} paclobutrazol to provide significant growth control after six weeks (Table 1). In contrast, petunia required 4 or 16 mg·L^{-1} paclobutrazol to control diameter and height, respectively (Table 1). Depending on the desired level of control, growers should use 4–8 mg·L^{-1} paclobutrazol for petunia, or ≥32 mg·L^{-1} paclobutrazol for dahlia. This recommendation for petunia is similar to the recommendation of 4–6 mg·L-1 paclobutrazol provided by Blanchard and Runkle (2007).

CONCLUSION

Pre-transplant liner soaks of as little as 4 mg·L^{-1} provided significant growth control in petunias, while dahlia required much higher paclobutrazol concentrations. This indicates that dahlia simply required greater paclobutrazol concentrations for significant growth control, or that paclobutrazol was only moderately effective in controlling dahlia growth. Other PGRs may be more effective for growth control in dahlia and should be tested in further studies investigating pre-transplant PGR liner soaks.

LITERATURE CITED

Blanchard, M.G. and E.S. Runkle. 2007. Dipping bedding plant liners in paclobutrazol or uniconazole inhibits subsequent stem extension. HortTechnology 17(2):178–182.

Latimer, J. and V. Groover. 2005. PGR liner dips on herbaceous perennials. Greenhouse Product News 15(9):38–44.

Schnelle, R., C. Cerveny, and J. Barrett. 2005. Factors affecting PGR liner dips. Greenhouse Product News 15(7):20-25.

Whipker, B.E., I. McCall, B. Krug, and M. Bell. 2003. Growth control of vegetative annuals with ancymidol, flurprimidol and paclobutrazol liner dips. Proceedings of the Plant Growth Regulation Society of America 30:198-202.

Figure 1. 'Dahlianova Hypnotica Purple Bicolor' dahlia plants grown for six weeks with pre-transplant paclobutrazol liner soaks.

Table 1. Mean heights and diameters for biweekly measurements of 'Dahlianova Hypnotica Purple Bicolor' dahlia and 'Crazytunia Razzmatazz' petunia.

Growth Parameter (cm)	Week	Paclobutrazol concentration (mg·L^{-1}) 0	4	8	16	32	64
Dahlia							
Height	2	6.8 g[1]	6.8 g	6.5 g	6.5 g	5.8 g	5.3 g
Height	4	16.9 c	16.1 cd	15.2 cde	14.0 def	13.1 ef	12.2 f
Height	6	30.2 a	28.4 a	29.4 a	28.5 a	25.6 b	23.5 b
Diameter	2	17.6 g	16.4 gh	14.2 hi	14.9 ghi	12.9 ij	10.6 j
Diameter	4	32.6 bcd	33.0 abc	31.1 cde	30.0 de	28.7 ef	26.7 f
Diameter	6	34.8 ab	35.5 a	34.7 ab	34.0 ab	33.4 abc	32.4 bcd
Petunia							
Height	2	7.1 efgh	3.7 h	5.1 gh	3.0 h	3.0 h	3.4 h
Height	4	15.3 abc	9.9 cdefg	11.4 cdef	8.2 defgh	7.7 efgh	6.1 fgh
Height	6	20.3 a	19.4 a	17.6 ab	12.5 bcde	13.5 bcd	12.3 bcde
Diameter	2	9.6 ef	7.0 f	7.2 f	5.6 f	6.5 f	6.5 f
Diameter	4	16.7 d	14.7 de	14.8 de	11.0 def	10.0 ef	8.5 ef
Diameter	6	38.8 a	27.6 b	23.5 bc	17.4 cd	16.4 d	16.6 d

[1] Lower case letters signify minimum significant differences among weeks of each growth parameter. Means with different letters were significantly different at $P \leq 0.05$.

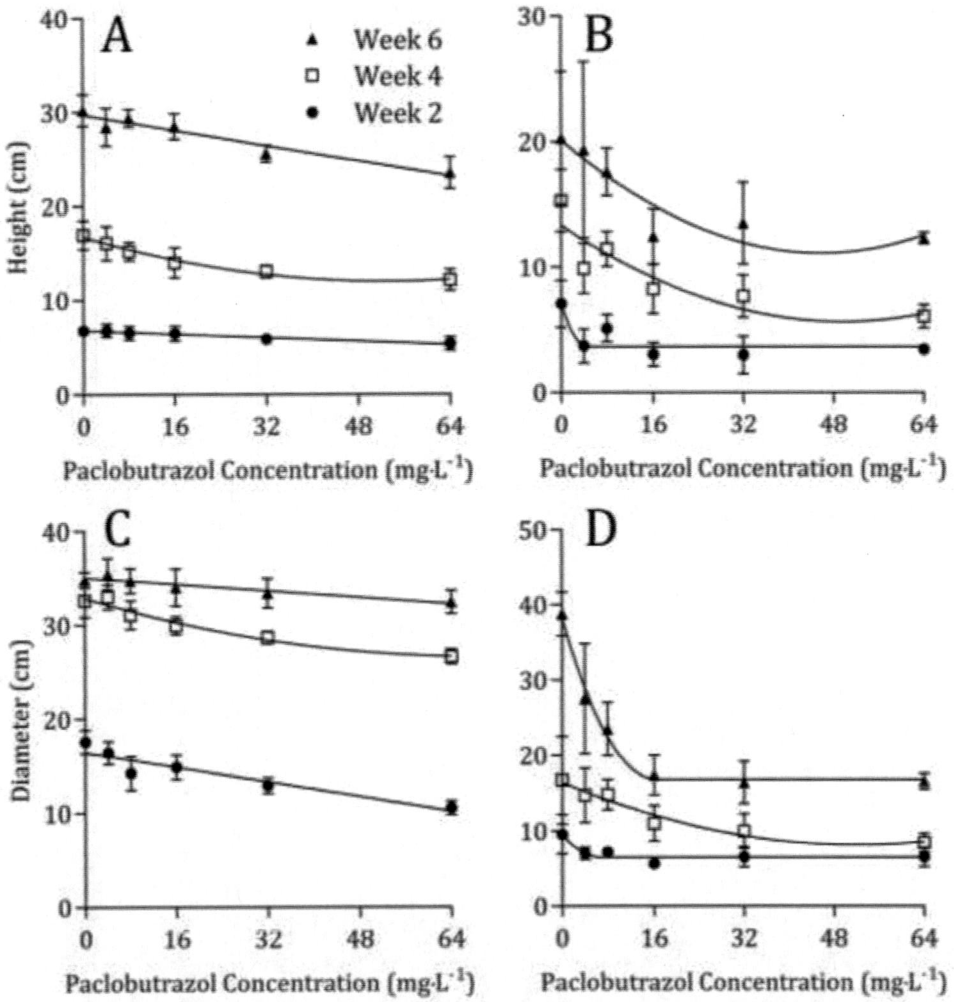

Figure 2. Best fit linear, quadratic, or quadratic plateau regression models for three biweekly measurements of (A) 'Dahlianova Hypnotica Purple Bicolor' dahlia height, (B) 'Crazytunia Razzmatazz' petunia height, (C) dahlia diameter, and (D) petunia diameter. Selected models significant at $P \leq 0.01$

Table 2. Best fit (Adj–r^2, AICc, DIF[1]) regression models for height and diameter of 'Dahlianova Hypnotica Purple Bicolor' dahlia and 'Crazytunia Razzmatazz' petunia.

Growth parameter	Week	Model[2]	Regression equation	Adj–r^2	AICc	DIF	X_0 / Vertex[3]
Dahlia							
Height	2	L	$6.76 - 0.0234x$	0.40	–	–	>64.0
Height	4	Q	$16.73 - 0.177x + 0.00167x^2$	0.61	25.2	5.8	53.0
Height	6	L	$29.71 - 0.101x$	0.68	–	–	>64.0
Diameter	2	L	$16.45 - 0.0977x$	0.70	–	–	>64.0
Diameter	4	Q	$32.84 - 0.180x + 0.00131x^2$	0.75	22.4	3.2	68.3
Diameter	6	L	$35.03 - 0.0437x$	0.30	–	–	>64.0
Petunia							
Height	2	QP	$7.05 - 1.472x + 0.1576x^2$	0.36	31.6	4.7	4.7
Height	4	Q	$13.33 - 0.311x + 0.00313x^2$	0.58	64.2	5.1	49.7
Height	6	Q	$20.15 - 0.395x + 0.00431x^2$	0.34	106.8	3.5	45.8
Diameter	2	QP	$9.57 - 0.897x + 0.0649x^2$	0.38	30.0	7.5	6.9
Diameter	4	Q	$16.41 - 0.371x + 0.00303x^2$	0.45	88.9	11.0	61.2
Diameter	6	QP	$34.06 - 1.078x + 0.01276x^2$	0.82	102.6	18.5	16.4

[1] Adj–r^2, adjusted coefficient of determination; AICc, corrected Akaike Information Criterion; DIF, difference between selected model AICc and inferior model AICc when considering quadratic and quadratic plateau models

[2] Models: L, linear; Q, quadratic; QP, quadratic plateau

[3] X_0 values where QP models plateau, and vertex values are minimum point on model. Both indicate the paclobutrazol concentration ($mg \cdot L^{-1}$) required to minimize growth.

Figure 3. 'Crazytunia Razzmatazz' petunia plants grown for six weeks with pre-transplant paclobutrazol liner soaks.

THE EVOLUTION OF THE USE OF PLANT GROWTH REGULATORS ON POINSETTIAS (EUPHORBIA PULCHERRIMA)

James E. Barrett*and Carolyn A. Bartuska
Environmental Horticulture Dept., University of Florida, Gainesville, Florida.

Poinsettias is a major crop in the international floriculture industry, and the evolving role of PGRs in the production of flowering crops was driven by the industry's understanding of the use of PGRs on poinsettias. The natural growth habit of poinsettias in the wild is a tall, poorly branched, small shrub not well suited for commercial container production. However, that was the style plant used in production until the 1970s. During that period and since there have been major advancements in breeding to improve plant growth habit and reduce sensitivity to ethylene during shipping. A significant increase in demand for poinsettias occurred with the marketing efforts of the Ecke company, who was the major poinsettia breeder in North America. They tightly connected poinsettias with the Christmas holidays in the minds of consumers. The next factor allowing for the rapid growth in poinsettia production was the ability of the growers to control plant size using Gibberellin biosynthesis inhibitors, which are often referred to as growth retardants. One of the earliest commercial uses of chlormequat chloride was on poinsettias. In the early 1980s the use of a tank-mix of chlormequat chloride and daminozide became important in warm climates and for more vigorous cultivars. Today paclobutrazol is the dominant PGR in ornamental production. During the 1990s and 2000s the industry learned to use a much more active chemical with a much greater demand for precession of methods of application, timing, and variations in optimum concentrations due to weather, cultivar, and cultural conditions. Learning to use paclobutrazol opened the industry to understanding more technical approaches to crop production and use of pesticides. This wide spread use of paclobutrazol and recent changes to recirculating irrigation water for environmental reasons has led to a major situation facing many commercial operations now is finding ways to remove paclobutrazol from recaptured irrigation water.

EXPANDING THE USE OF LIGHT TO DIRECT PLANT GROWTH

Jake Holley, Shun Nishimura, Jake Hubert, Rachel Schuster, and Melanie Yelton
LumiGrow Inc., 1480 64th St., Suite 150, Emeryville, CA 94608

At LumiGrow we create lighting strategies to optimize plant production. We have developed a partnership approach to work with Universities and growers to complement our in-house basic research program. We will discuss recent progress in three areas.

First, in growth chambers we determined that a photoperiod of 21 hours is optimum, across several species, for plant height and biomass. We are adapting this work to the greenhouse environment and extending the work to fruiting plants. As a component the work we found with red lettuce the photoperiod which produces the greatest biomass, is not necessarily the best photoperiod for color development. To optimize lighting strategies, photoperiods and intensities can be adjusted to regulate plant growth, depending on whether the goal is for color or biomass production.

Also, light continues to show promise as a valuable tool to mitigate plant disease. A new study demonstrated that night interruption with LED lights can be used to reduce basil downy mildew infestation. Also, in a large production environment powdery mildew was reduced in cucumbers grown under LED lighting, when compared to HPS lighting. For cucumber, lighting appeared to aid in the regulation of disease resistance within leaves.

Second, we have extended this partnership approach to commercial greenhouse trials to determine the optimum light levels that integrate with the farmer's economic objectives. We will present strawberry production as a model of this approach. Additional trials with spectrum will be used to impact flavor, growth habit, and postharvest shelf-life of the berries.

Finally, we are using new software to create improved lighting plans. These new plans often require the lights to be hung in unexpected patterns to meet uniformity and consistency parameters. To achieve more consistent lighting within the greenhouse, we are also trialing a light sensor which measures ambient light and adjust supplemental lighting to meet the target light level required for standardize growth and production. This will tremendously aid in the application and consistency of using light as a plant growth regulator.

PLANT GROWTH STIMULATOR TRIALS: TRIAL DESIGN, EVALUATION, CONDUCT AND THE INEVITABILITY OF FAILURE—EXPERIENCE MAKES YOU BETTER

Steve West
Research Designed for Agriculture

Finding methods to realistically evaluate inputs which are intended to stimulate or affect plant growth in a positive way are very difficult to accomplish in field settings. One must relentlessly work at developing and testing methods and techniques to tease out subtle differences and determine the circumstances when a product may have an effect. The challenge of dreaming up how to quantify "happy" plants is a daily undertaking. Adding the nearly infinite variables of a field ecosystem complicates everything. The reality of finite resources such as qualified observers, effective equipment and technology, and, of course, time, further confounds efforts. Being observant, nimble and persistent in finding ways to evaluate plant happiness is the cornerstone to success. Planning, trial design, and excellence in execution of husbandry to limit variables is critical as a foundation. One must be agile in methodologies, and follow intuition when observations, data, imagery, etc., reveals new paths of exploration. These paths will point to new research directions, or, better yet, marketable components to a technology. This talk will look at the beginnings of how to evaluate plant happiness in field situations—with a healthy discussion of failures. Because experience does make you better.

INVESTIGATION OF THE OPTIMUM PLOT SIZE FOR ACCURATE YIELD EVALUATIONS IN AGRICULTURE

Marcus Jones[1], Ken Moore[2], Mark Risley[1], Catherine White[1]
[1]WinField United, Shoreview, MN 55126
[2]Department of Agronomy, Iowa State University, Ames, IA 50011

Field trials investigating yield benefits of agricultural practices or inputs utilize randomization, replication, and blocking as fundamentals of statistical design. Plot size is an overlooked design fundamental that could influence the outcome of trials as fields often are not uniform and tend to be unpredictable in gradients (Jones et al., 2015).

Yield based field trials routinely utilize plots 3.048 m wide and plot lengths from 5.3 to 15.2 m in length. Previous research suggests that statistical metrics of crop yields can be influenced by plot size characteristics. Wassom and Kalton (1953) discovered small plots 1.0 x 2.2 m in size were most efficient at estimating dry matter of bromegrass. Plots 229 m or greater in length provided much less experimental error in wheat and barley yield trials (Wuest et al., 1994). Masood and Raza (2012) found that the coefficient of variation of rice yields decreased as plot size increased and the decrease was maximized with plots square in shape. Further, Tedford et al. (2016) demonstrated that yield benefits from strobilurin fungicides applied to maize increased as plot size increased. To better understand the effect of plot shape and size on data variability measured as maize yield, a series of uniformity trials were conducted across the upper Midwest during the 2015-2016 cropping seasons.

A uniformity trial is a type of experiment where all cultural and management practices are applied uniformly to a test area to minimize sources of variability except those due to native soil differences (Gomez and Gomez, 1984). The area is then divided into individual plots and yield measurements are recorded. The inherent variability of the soil can be measured by examining differences among plots.

Field trials were conducted in 2015 and 2016 at the Iowa State University Agronomy Research Farm near Boone, IA and near River Falls, WI on a private research site operated by WinField United. A description of the soil, planting population, and previous crop for each site is included in Table 1. An area roughly 0.3 hectare was bulk planted with a single maize hybrid and divided into 144 experimental units with border on all sides. Individual plots measured 3.0 m wide and 4.6 m in length with 0.76 m spacing. The middle two rows were harvested, and yield calculated.

Prior to harvest, the plot length of the middle two harvest rows was measured to the nearest 12.7 mm and combine yields recalculated based on the actual plot length. Stand counts and pre-harvest ear counts were collected as quality control measures to ensure differences in plot yields were due to plot characteristics and not other factors. Statistical variance for 31 different plot combinations ranging in size from 27.9 m^2 to 1,003 m^2 was calculated by:

$$\frac{\Sigma(x-\bar{x})^2}{(n-1)}$$

Trial average maize yield ranged from 13235.0 to 18023.3 kg ha^{-1} and the range in yield was 5037.1-7767.5 kg ha^{-1} (Table 1). Statistical variance decreased as plot size increased (Figure 1). Variance was reduced over 70% with plots larger than 83 m^2 using a 27.9 m^2 size plot as comparison. When plot size was held constant, the shape of the plot did not consistently lower variance (data not shown).

The data also illustrate the relationship between plot size and replicate number (Figure 2). There is a tradeoff between these two design fundamentals and fewer replicates are needed as plot size increases since larger plots better capture field variability. From a statistical perspective, a 148.6 m^2 plot with two replicates will produce an equivalent least significant difference (LSD) compared to a 27.9 m^2 plot with five replicates. While increasing replicates to small plots (27.9 m^2) can also be an effective strategy to reduce variance, the benefit decreases with each additional replicate and replicates fail to address edge and alley effects.

Effects from borders and alleys have been shown to elevate crop yields of plants near the edge of plots (Geater et al., 2004; Wang et al., 2013; Vincelli and Lee, 2015). This is likely due to less competition and greater availability of resources. Since most plants in small plots (27.9 m^2) are not far away from an alley they are likely near their maximum yield potential. Because larger plots reduce the perimeter to area ratio and a greater percentage of plants are away from alleys, per plant yields could be lower allowing more room for yield enhancement from agronomic inputs. The yield advantage of strobilurin fungicides applied to corn increased from 403.5 kg ha^{-1} in small plot trials up to 1210.5 kg ha^{-1} when evaluated in commercial strip trials (Tedford et al., 2016).

These results demonstrate that the variation in small plot trials is not representative of conditions in larger bulk planted fields. Larger plot sizes may also help improve data quality by reducing the influence of drift from neighboring plots and adjacent experiments. Plot size can be used to manage field variability and should be considered along with replication and randomization when planning a yield trial.

REFERENCES

Geater, C., Jones, M., Burger, B., 2004. Understanding the effects of experimental design on experimental error in maize hybrid evaluation. In: Proceedings of the 59th Annual Corn and Sorghum Seed Research Conference. American Seed Trade Association, Alexandria, VA. CD-ROM.

Gomez, K.A. and A.A. Gomez, (1984). Statistical procedures for agricultural research (2 ed.). John wiley and sons, NewYork, 680 p.

Jones, M., R. Woodward, and J. Stoller. 2015. Increasing precision in agronomic field trials using latin square designs. Agron. J. 107:20-24. doi:10.2134/agronj14.0232.

Masood M. A., and Raza I. Estimation of optimum field plot size and shape in paddy yield trial. 2012. American-Eurasian Journal of Scientific Research.7:264-269.

Tedford, E., Kriss, A., Geater, C., Saini, M., Battles, B., Smelser, R., Fithian, W. 2017. Plot size can influence yield benefits from fungicides on corn. Crop Protection 91:66-73.

Vincelli, P., Lee, C., 2015. Influence of open alleys in field trials assessing yield effects from fungicides in corn. Plant Dis. 99: 263-266.

Wang, K., Zhou, H., Wang, B., Jian, Z., Wang, F., Huang, J., Nie, L., Cui, K., Peng, S., 2013. Quantification of border effect on grain yield measurement of hybrid rice. Field Crops Res. 141:47-54.

Wassom, C. E. and Kalton, R. R. 1953.) Estimations of optimum plot size using data from bromegrass uniformity trials." Research Bulletin (Iowa Agriculture and Home Economics Experiment Station): Vol. 31: No. 396, Article 1.

Wuest, S.B., B.C. Miller, I.R. Alldredge, S.O. Guy, R.S. Karow, R.J. Veseth, and D.J. Wysocki. 1994. Increasing plot length reduces experimental error of on-farm Tests. I. Prod. Agric. 7:211-215.

Table 1. Agronomic and summary statistics of six uniformity trials conducted to investigate the effect of plot shape and size on the variance of maize yield.

Trial information	2015			2016		
Trial host	ISU B1	ISU B4/B5	Wahr1	ISU NW	ISU SE	Wahr2
Previous crop	Corn	Corn	Corn	Soybean	Corn	Soybean
Plants/hectare	79,073	79,073	74,131	75,614	78,826	79,073
Soil	Nicollet	Nicollet	Pillot	Canisteo & Harps	Canisteo & Clarion	Pillot
Summary statistics						
Yield (kg ha^{-1})	15514.8	14559.9	13235.0	18023.3	17209.5	16597.6
Std Dev	18.6	17.9	20.8	16.0	16.8	14.2
Minimum	10188.5	10834.1	8184.5	15467.7	12549.0	13504.0
Maximum	17599.6	17108.7	15952.0	20504.9	19684.4	20067.7
Range	7411.1	6274.5	7767.5	5037.1	7135.3	6563.7
Count	144	144	192	144	144	144

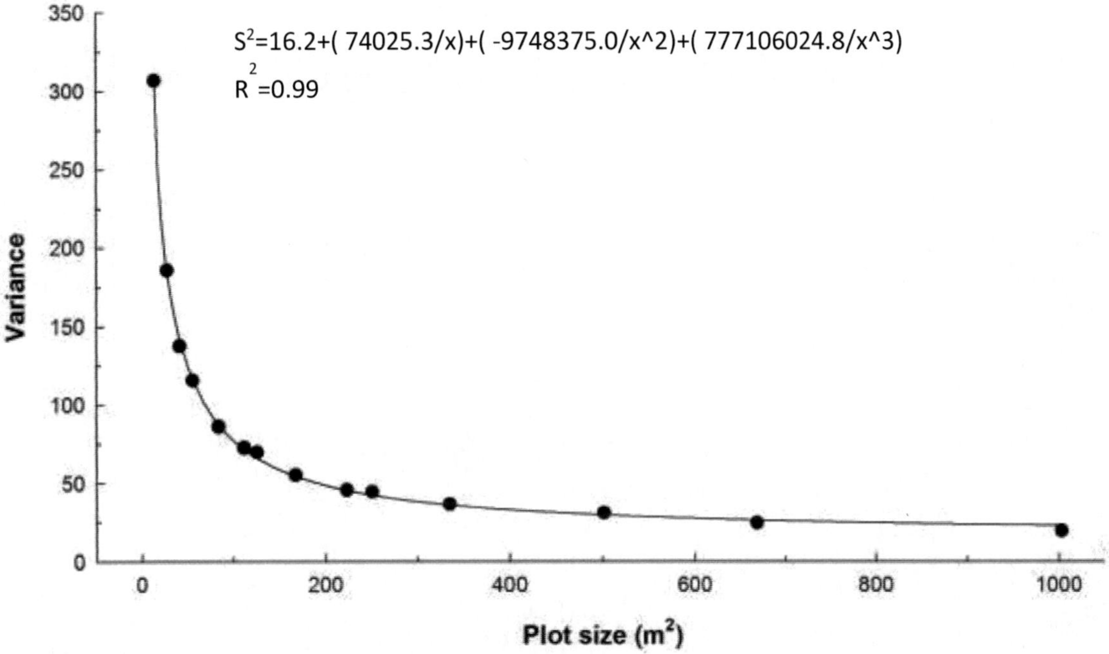

Figure 1. Trial variance decreases as plot size increases.

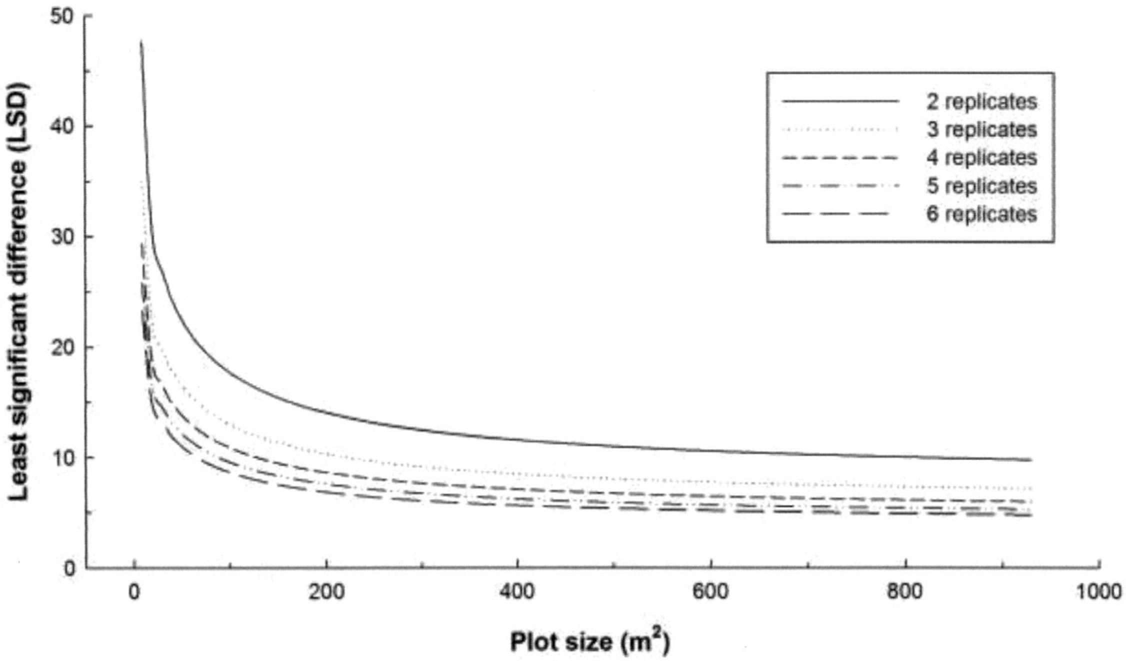

Figure 2. A graphical representation of the tradeoff between plot size and number of replicates.

EXPLORING PLANT GROWTH REGULATORS TO REDUCE BOLTING IN CILANTRO GROWN IN GREENHOUSE CONDITIONS AND OPEN FIELD PRODUCTION

Ute Albrecht, Bryce Meyering, Monica Ozores-Hampton

University of Florida/IFAS, Southwest Florida Research and Education Center, Immokalee, FL 34142

Cilantro (Coriandrum sativum), also known as Chinese parsley, Mexican parsley, or coriander is a cool-season vegetable and is grown in many regions throughout the US. The optimal temperatures of growth are between 10 and 28 °C and it will bolt easily in hot and long-day conditions. Bolting renders the crop unmarketable and poses an economic problem to specialty growers not only in Florida, but also in other production areas.

In long-day biennial plants bolting and flowering initiation is usually promoted by gibberellins, although the mechanisms vary between species. This study investigates whether inhibitors of gibberellic acid (GA) synthesis can inhibit or delay bolting in cilantro 'Santos', grown in commercial conditions in Florida. Preliminary greenhouse studies using inhibitors of GA that act early in the GA synthesis pathway, such as paclobutrazol (PBZ) and chlormequat-chloride (CCC) and a late GA inhibitor, prohexadione calcium (ProCa), demonstrated an up to 8-fold reduction in bolting compared with untreated plants, whereas GA promoted bolting. In addition to the greenhouse studies, three field trials were conducted in Clewiston, FL, in a commercial farm between November 2016 and April 2017. In trial 1 (Nov 2016-Jan 2017), treatments were applied 6 weeks after seeding (WAS) and plants were assessed 4 weeks after treatment application (WAT). In trial 2 and 3 (Feb 2017-Apr 2017) treatments were applied at 5 and 6 WAS, respectively, and plants were assessed 3.5 to 4 WAT. Products were applied foliar, except for PBZ, which was applied to the root zone, at rates recommended by the manufacturers. In trial 1, the lowest bolting (25%) based on assessment of internode length was observed for plants treated with ProCa compared with plants treated with PBZ and untreated plants, which bolted between 68% and 98%, respectively. The percentage of plants with a visible flower stalk was 48% for untreated plants, 22% for PBZ, and 2% for ProCa. In trial 2, bolting did not exceed 5% in any of the treatments at time of assessment. In trial 3, bolting was 9% for ProCa, 14% for PBZ, and 40% for untreated plants. The difference in overall bolting between trial 2 and trial 3 may be related to a sudden drop in temperature few days after germination of plants in trial 2. After bolting assessment in trial 2 and 3, half of the plants in each treatment plot were cut at 2-3 cm above the ground and the other half was left uncut. In trial 2, assessment of bolting 1 and 2 weeks thereafter did not yield any differences among treatments. In trial 3, the number of bolting plants was 2 to 3-fold lower in PBZ and ProCa compared with untreated plants at 11 WAS, whether plants were cut or uncut. Incidence of bolting was not significantly different between treatments at 12 WAS in both cut and uncut plants. Additional studies are necessary to clearly define the best time of application to reduce bolting and to determine which type of growth regulator is most effective.

PRODUCTION TECHNOLOGIES OF MYCORRHIZAE: A GLOBAL COMPARISON AND FUTURE POTENTIAL

Alok Adholeya
Mycorrhizal Applications LLC

The importance of mycorrhizae is well established for sustainable and more productive agriculture and horticulture activities worldwide. The commercial scale production of mycorrhizae, particularly Arbuscular Mycorrhizal fungi, has been a challenge until recently. Various innovative technologies and methods were adopted to develop successful commercial production. Technologies such as Greenhouse in vivo, Hydroponics, nutrient film technique, Aeroponics and root organ culture methods were attempted. Success and failures, advantages and limitations will be discussed. Application of mycorrhizae in parallel has yielded many successful case studies which allowed global farming communities to adopt mycorrhizal application within their package of practices. A booming success is around the corner towards agriculture sustainability, resource use efficiency and economic benefits.

USE OF ARBUSCULAR MYCORRHIZAL FUNGI TO IMPROVE SALINITY STRESS TOLERANCE OF TRANSPLANTED CITRUS TREES

Jozsef Racsko, Alok Adholeya, Stephanie Garcia
Mycorrhizal Applications, LLC. 710 NW E. Street, Grants Pass, OR 97526

Application of mycorrhizal inocula in agricultural systems has become increasingly popular to improve crop performance. The effectiveness of inoculum in agro-ecosystems must be assessed for the continued development of these inoculum production technologies. The goal of this study was to investigate the efficacy of MycoApply® inoculum in container-grown citrus trees. In spring of 2016, commercially available MycoApply® arbuscular mycorrhizal inoculum containing four arbuscular mycorrhizal species (Rhizophagus intraradices, Funneliformis mosseae, Claroideoglomus etunicatum, Glomus aggregatum) (trade name MycoApply Endo) were applied to 'Meyer' lemon trees at a rate of 6,000 propagules per tree. Single trees were planted into 450 kg tote bags filled with granular clay medium and grown under severe salinity stress (100 mM NaCl drenched weekly). ANOVA analysis was used to determine statistical significance at a P level < 0.05. MycoApply Endo inoculum were effective in mitigating salinity stress effects and increasing mycorrhizal colonization, vegetative growth, relative chlorophyll content, yield, fruit size, photosynthesis, and chlorophyll fluorescence. Differences in the nutrient content were observed between the leaves of inoculated trees and the untreated trees. Mineral nutrient content in the leaves increased for N, P, K, Mn, Zn and decreased for Ca, B, Al and Cu. This study confirmed the efficacy of arbuscular mycorrhizal fungi to improve performance and salinity stress tolerance of citrus trees as host plant.

PHYTOHORMONE SIGNALLING IN ARBUSCULAR MYCORRHIZA DEVELOPMENT

Caroline Gutjahr
Plant Genetics, School of Life Science Weihenstephan, Technical University of Munich (TUM), Emil Ramann Str. 4, 85354 Freising, Germany
Email: caroline.gutjahr@tum.de

Arbuscular mycorrhiza (AM) is a symbiosis between fungi of the glomeromycota and about 80% of land plant species (Smith and Read, 2008). The fungi colonize the interior of plant roots and form extended and branched extraradical hyphal mycelia in the soil. With these hyphae they capture soil mineral nutrients, transport them into the root and release them at highly branched hyphal structures, the arbuscules, which form inside cortex cells (Smith et al., 2011). In addition to nutritional benefits, AM symbiosis also increases biotic and abiotic stress resistance of plants. For these reasons, there is increasing interest in application of AM in sustainable agricultural practices to maintain yield while reducing mineral fertilizer and to buffer plant performance against environmental variability (Sawers et al., 2008a; Sawers et al., 2008; Gianinazzi et al., 2010).

The main research interest in my laboratory is the developmental biology of AM formation, with a special focus on the role of plant hormones in AM development. Previously, we have found that the karrikin receptor complex is required for AM formation in rice (Gutjahr et al., 2015). Karrikin is a butenolide compound found in smoke, which induces germination of fire following plants (Waters et al., 2013). It is perceived by the alpha/beta-fold hydrolase D14L/KAI2, which forms a complex with the F-box protein D3/MAX2 (Waters et al., 2017). Rice mutants of both D14L and D3 do not allow root colonization by AM fungi and colonization is blocked prior to or at hyphopodium formation, the earliest stage of AM development (Gutjahr et al., 2015). Hyphopodia are lentil shaped hyphal structure, that the fungi develop to attach to the root surface. In addition, the d14l mutant is impaired in the transcriptional response to exudates from germinating fungal spores, indicating that the karrikin perception machinery is involved in pre-contact chemical signaling between plant and fungus (Gutjahr et al., 2015). We are now elucidating the molecular mechanisms downstream of the karrikin receptor complex, which support AM development in the model legume Lotus japonicus.

It has long been known that gibberellins (GAs) inhibit AM formation, when applied to roots inoculated by AM fungi (El Ghachtouli et al., 1996) and it has been found that DELLA proteins, the repressors of GA signaling are required for the formation of arbuscules (Floss et al., 2013; Foo et al.,

2013). We have identified a molecular mechanism that may explain how DELLAs act in arbuscule formation (Pimprikar et al., 2016). DELLAs interact with CYCLOPS, a central transcriptional regulator of AM development and nodulation (Yano et al., 2008; Singh et al., 2014; Jin et al., 2016; Pimprikar et al., 2016). CYCLOPS – likely in complex with DELLA and the calcium calmodulin-dependent kinase CCaMK - binds to the promoter of RAM1 (REDUCED ARBUSUCLAR MYCORRHIZA1), a GRAS transcription factor, which is required for arbuscule branching (Pimprikar et al., 2016). Thus, DELLA participates in a transcriptional cascade, which regulates arbuscule formation. The complex of CYCLOPS and DELLA appears to be a signaling hub in which the plant can directly connect symbiosis and hormone signaling. This may enable the plant to orchestrate the amount of root colonization, with its physiological state under changing environmental conditions.

In addition to investigating the molecular basis of arbuscular mycorrhiza development we are interested in application of AM in agriculture. It has been observed that positive growth responses (AM responsiveness) differs among cultivars of ecotypes of the same species (reviewed in (Sawers et al., 2008)). This indicates that there are likely genetic polymorphisms underlying the differential performance response to AM. Finding these polymorphisms would be of high interest for breeding AM-optimized crops. In the frame of the BayKlimaFit project (http://www.bayklimafit.de) we have performed a screen with maize inbred lines of the European Dent breeding pool for AM-mediated drought stress resistance. To this end we grew plants in a rain out shelter with minimal watering. In one field trial in summer 2016 we found three maize lines, which profited from inoculation with Rhizophagus irregularis with increased yield, when grown in the rain out shelter under artificial drought conditions (Muth and Gutjahr, unpublished). We are currently investigating whether these results are reproducible in another field season. This would be highly interesting as a basis for QTL-mapping of genetic loci, related to AM-mediated drought resistance increase.

REFERENCES

El Ghachtouli, N., Martin-Tanguy, J., Paynot, M., and Gianinazzi, S. (1996). First-report of the inhibition of arbuscular mycorrhizal infection of Pisum sativum by specific and irreversible inhibition of polyamine biosynthesis or by gibberellic acid treatment. FEBS letters 385, 189-192.

Floss, D.S., Levy, J.G., Lévesque-Tremblay, V., Pumplin, N., and Harrison, M.J. (2013). DELLA proteins regulate arbuscule formation in arbuscular mycorrhizal symbiosis. Proc. Natl. Acad. Sci. USA 110, E5025-E5034.

Foo, E., Ross, J.J., Jones, W.T., and Reid, J.B. (2013). Plant hormones in arbuscular mycorrhizal symbioses: an emerging role for gibberellins. Ann Bot 111, 769-779.

Gianinazzi, S., Gollotte, A., Binet, M.-N., Tuinen, D., Redecker, D., and Wipf, D. (2010). Agroecology: the key role of arbuscular mycorrhizas in ecosystem services. Mycorrhiza 20, 519-530.

Gutjahr, C., Gobbato, E., Choi, J., Riemann, M., Johnston, M.G., Summers, W., Carbonnel, S., Mansfield, C., Yang, S.-Y., Nadal, M., Acosta, I., Takano, M., Jiao, W.-B., Schneeberger, K., Kelly, K.A., and Paszkowski, U. (2015). Rice perception of symbiotic arbuscular mycorrhizal fungi requires the karrikin receptor complex. Science 350, 1521-1524.

Jin, Y., Liu, H., Luo, D., Yu, N., Dong, W., Wang, C., Zhang, X., Dai, H., Yang, J., and Wang, E. (2016). DELLA proteins are common components of symbiotic rhizobial and mycorrhizal signalling pathways. Nature Communications 7, 12433.

Pimprikar, P., Carbonnel, S., Paries, M., Katzer, K., Klingl, V., Bohmer, M., Karl, L., Floss, D., Harrison, M., Parniske, M., and Gutjahr, C. (2016). A CCaMK-CYCLOPS-DELLA complex regulates transcription of RAM1, a central regulator of arbuscule branching. Curr Biol 26, 987-998.

Sawers, R., Gutjahr, C., and Paszkowski, U. (2008). Cereal mycorrhiza: an ancient symbiosis in modern agriculture. Trends Plant Sci 13, 93 - 97.

Singh, S., Katzer, K., Lambert, J., Cerri, M., and Parniske, M. (2014). CYCLOPS, a DNA-binding transcriptional activator, orchestrates symbiotic root nodule development. Cell Host Microbe 15, 139-152.

Smith, S., and Read, D. (2008). Mycorrhizal Symbiosis. (Academic Press London).

Smith, S.E., Jakobsen, I., GrÃ¸nlund, M., and Smith, F.A. (2011). Roles of arbuscular mycorrhizas in plant phosphorus nutrition: interactions between pathways of phosphorus uptake in arbuscular mycorrhizal roots have important implications for understanding and manipulating plant phosphorus acquisition. Plant Physiol 156, 1050-1057.

Waters, M.T., Scaffidi, A., Flematti, G.R., and Smith, S.M. (2013). The origins and mechanisms of karrikin signalling. Curr Opin Plant Biol 16, 667-673.

Waters, M.T., Gutjahr, C., Bennett, T., and Nelson, D.C. (2017). Strigolactone Signaling and Evolution. Ann Rev Plant Biol 68, 291-322.

Yano, K., Yoshida, S., Müller, J., Singh, S., Banba, M., Vickers, K., Markmann, K., White, C., Schuller, B., Sato, S., Asamizu, E., Tabata, S., Murooka, Y., Perry, J., Wang, T., Kawaguchi, M., Imaizumi-Anraku, H., Hayashi, M., and Parniske, M. (2008). CYCLOPS, a mediator of symbiotic intracellular accommodation. Proc. Natl. Acad. Sci. USA 105, 20540-20545.

REGISTERING PLANT GROWTH REGULATORS: GLOBAL POLICIES WORLDWIDE AND HOW TO WORK WITHIN THEM

Maria Herrero
Valent BioSciences LLC

Commercial use of plant growth regulators (PGRs), whether naturally occurring or synthetic, must undergo an evaluation and registration by authorities in almost all countries of the world. Policies regulating their definition, development and registration vary from country to country. Regulators suggest that PGRs and pest control products must be fully and comprehensively evaluated through a probabilistic toxicology and environmental risk assessment process, with proven efficacy in controlled field studies. Yet, a lack of a uniform regulatory model adds to the complexity of bringing PGRs to the forefront in use and to the concept of using them in modern integrated pest management strategies. Proof of natural occurrence, or evidence that the compounds are structurally-similar and functionally identical to naturally-occurring compound permits a tiered evaluation process in many countries. For naturally occurring compounds, literature on metabolism, historical exposure by humans, non-targets, beneficials, soils micro-organisms, etc. may prove evidence of scientific justifications for not conduction specific studies even in countries that do accept a tiered evaluation system. A review will be given of regulatory study requirements by areas of the world. Constant attention must be paid as regulation continues to change worldwide and evaluation requirements increasing in scope.

MINERAL-BASED MODULATION OF PGR POOLS IN KIWIFRUIT IMPROVES PRODUCTION AND YIELD METRICS OVER CYTOKININ OXIDASE INHIBITORS.

Richard Woodward and Jerry Stoller

StollerUSA, Stoller Group, Houston, TX 77024

Plants respond to signals from their environment with the production growth regulators (PGR) and signaling molecules. Environmental conditions define the pool size of these growth regulators and signals by effecting biosynthesis, degradation, conjugation, transport and perception. Phrenological stage determines temporal deployment of specific growth regulators and signaling molecules. Crosstalk within the active PGR / signal pool defines plant health, synchrony, productivity, fruit quality and ultimately yield. We examined the role minerals play in controlling the pool size of regulating molecules in Kiwifruit as manifested by synchrony, fruit quality [solids, grade] and yield metrics in case studies at three locations covering up to five growing seasons. We compared the cytokinin oxidase inhibitor, Forchlorfenuron (CPPU) alone, to CPPU in combination with a more robust mixture of actual PGRs [cytokinin, auxin (IBA) and gibberellic acid (GA3)] and supportive minerals boron and calcium. Four sequential, annual trials conducted from 2013 through 2016 demonstrated viable alternatives to CPPU in fruit sizing, quality and yield. We evaluated fruit set on artificially pollinated plants treated before flowering, with a foliar application of minerals that manage critical PGR pools. Supporting minerals improved the fruit set of artificially pollinated flowers by 26%. Mineral support for supplemented or endogenous PGR pools significantly reduced premature fruit drop and increased soluble solids and fruit grade size. We managed ethylene in mineral-modulated PGR programs to promote synchrony, accelerate bud break, and improve fruit count, fruit grade and total yield. Collectively, these mineral-based PGR management and support programs stand as an effective, viable and cost-effective alternative to phenyl-urea derived, cytokinin oxidase inhibitors for improved production and quality.

CITRUS ROOTSTOCKS EXHIBIT DIFFERENT SCION GROWTH RESPONSES TO PRE-BUDDING DEFICIT IRRIGATION

Gurreet Brar[1] and Timothy Gast[2]

[1]Assistant Professor, California State University, Fresno, CA

[2]Biological Scientist, SWFREC, University of Florida, Immokalee, FL

A number of new rootstocks have been developed and adopted by Florida citrus nursery industry in recent years. Preliminary studies and nurserymen observations reveal that some of them respond to different cultural practices differently, especially irrigation. Therefore, the nursery production practices need to be adapted according to their water relations especially in terms of bud push and scion growth. This study was conducted to elucidate the differences in bud push and scion growth responses of 3 different rootstocks under three irrigation regimes. The experiments were conducted in citrus screenhouses at the University of Florida Southwest Florida Research and Education Center in Immokalee, FL. Three Trifoliate orange parentage rootstock were chosen: Swingle citrumelo(Citrus paradisi x Poncirus trifoliata), US-802 (Citrus grandis x Poncirus trifoliata) and X-639 (Citrus reshni x Poncirus trifoliata). The trees were acquired from a commercial citrus nursery in South Florida and were transplanted to 2.65 L Citrapots on arrival. After acclimation for two weeks, the trees were subjected to these different irrigation treatments: Well-Watered Control (irrigated to replace 100% ETc); Water stressed (irrigated to replace only 50% ETc) and Deficit Irrigation (Started off with 100%, changed to 50% two weeks prior to budding and then changed back to 100% ETc just after budding). Budding was performed using T-budding method by professional budders from a commercial nursery. Other cultural practices including fertilizers etc. were done as per commercial standards in Florida citrus nurseries. Buds were forced by bending and tying the rootstock and unwrapping was done 21 days after budding. In each rootstock-irrigation treatment combination, half the trees received Benzyl adenine (BA; Configure, Fine Americas) application on the buds just after unwrapping and the other half didn't receive any BA application. Fifteen replicate trees were used in each of the 18 treatment combination. Mid-day stem water potential (SWP) of the trees was measured periodically throughout the experimental period. Data were collected on bud survival, scion length, number of leaves, scion dry weight, root dry weight and final leaf area. Net photosynthesis and stomatal conductance were measured at three instances during the experimental period. Data were analyzed by three-way ANOVA using SPSS statistical package. Rootstocks and irrigation factors showed significant differences in total dry weight, scion length and number of leaves while percent bud survival was not significantly different between any treatment combinations. While benzyl adenine application on the buds did not show any difference in any of the other parameters observed, total scion length showed strong interaction of BA application with irrigation. Similar observation was reported in a previous study where drought stressed trees responded to application of exogenous BA in terms of scion growth as compared to no BA application. From the current study, it was found that these rootstocks (Swingle citrumelo, US-802 and X-639) showed significant differences in all measured parameters under different deficit irrigation regimes.

GLOBAL TRANSCRIPTOMIC ANALYSIS OF HYDROGEN CYANAMIDE-TREATED PEACH TREES

Tripti Vashisth[1], S. Chhajed, T. Colquhoun[2], J. Olmstead[3] and M. Olmstead[3]
[1]Horticultural Sciences Department, University of Florida/Institute of Food and Agricultural Research, Citrus Research and Education Center, Lake Alfred, FL
[2]Environmental Horticultural Department, University of Florida/Institute of Food and Agricultural Research, Gainesville, FL
[3]Horticultural Sciences Department, University of Florida/Institute of Food and Agricultural Research, Gainesville, FL

Peaches (Prunus persica L.) are of considerable interest as an alternative crop for growers in Florida who are looking to diversify from citrus. The temperate origin of peach present significant production challenges for subtropical areas. Most significantly, inadequate chill accumulation is a common occurrence in production areas seeking to take advantage of high value, early-market windows. Low-chill peach varieties grown in Florida, USA that do not receive adequate chill exhibit extended bloom periods (>2 weeks in length). Most of Florida's chill hours are accumulated in December and January, but in recent years accumulated chilling units continue to decline. Hydrogen cyanamide (HCN) is widely used around the globe where mild winters prevail and where there are insufficient accumulated chill units to coordinate budbreak, induce uniform leaf emergence, and compress the bloom period. However, many growers have induced significant crop injury because the timing for different fruit crops is not well defined. Additionally, the molecular mechanism of how HCN works is still poorly understood and therefore it is challenging to develop strategies for proper and effective use HCN on fruit crops in absence of chill hour accumulation and erratic weather. In order to understand the transcriptional pathways associated with HCN-induced dormancy release in peaches we analyzed the vegetative and reproductive buds of peach. The objective of this study was to examine how HCN alters the trancriptome especially genes associated with endodormancy and budbreak in peach floral and vegetative buds grown in subtropical climates. On December 19, 2015, when pollen grains were translucent, low chill cultivar 'TropicBeauty' trees were sprayed with 1.2% HCN (BudPro) and control trees were not sprayed. Vegetative and floral buds were collected from 4 branches per tree for 4 biological replicates before application and then 1, 3 and 7 days after application. RNA isolation was performed to carry out RNAseq analysis. HCN treated plants had both early and uniform bloom, vegetative bud break was visible approximately 30 days after HCN application. Whereas untreated control trees displayed a prolonged bloom and uneven leaf emergence suggesting HCN application helped in breaking the dormancy and synchronizing the bud break. Genes related to dormancy release such as dormancy-associated MADS-box transcription factor, flower bud initiation, and oxidation stress were found to altered in response to HCN treatment. Altogether our field and RNAseq data shows that HCN application can induce early and uniform bud break potentially by substituting for chill hours required for release of endodormancy and the transcriptional pathway is similar to innate endodormancy release in peach.

BLOSSOM THINNING OF STONEFRUIT AND APPLES USING ORGANOSILICONE SURFACTANTS WITH ETHEPHON

Joshua D. Klein[1] and Shlomo Cohen[2]

[1]ARO - Volcani Center, Rishon LeZion Israel

[2]Institute for Agriculture according to the Torah, Yad Binyamin Israel

Hand-thinning is often required for temperate-zone tree fruits to achieve commercial size, but it is one of the most expensive aspects of fruit production. We investigated spray application of organosilicone surfactants as blossom thinners for almonds, apricots, peaches, nectarines, and apples. Silwet-408 (commercially marketed by Adama as "Dilwet"), at 0.5% (v/v) or at 0.25% in combination with 100 ppm ethephon, was an effective bloom thinner, resulting in fruit with enhanced size, color, and total soluble solids. Flower parts (petals and carpel/receptacle) treated with effective concentrations of Dilwet had enhanced membrane leakage, while pollen grains from such flowers had decreased viability. The cost of materials and spraying is recouped by the enhanced value of the larger fruit and by the savings on hand labour.

S-ABSCISIC ACID (S-ABA) TREATMENTS CONTROL BITTER PIT IN HONEYCRISP APPLES

Poliana Francescatto[1], Terence Robinson[1], Jaume Lordan Sanahuja[1], Jozsef Racsko[2]
[1]Horticulture Section, Cornell University, Geneva, NY 14456
[2]Mycorrhizal Applications, LLC., Grants Pass, OR 97526

Many environmental conditions, crop status and mineral factors, such as drought and salinity, relative humidity, high light and temperatures, light crops and large fruit, and mineral imbalance conditions (excessive soil levels of nitrogen and potassium) have been shown to affect whole plant calcium uptake and translocation to the fruit, and consequently fruit susceptibility to calcium deficiency disorders. Calcium (Ca) uptake into fruit and leaves is dependent on xylemic water movement, and hence presumably driven by transpiration and stomatal activity and growth. Due to high transpiration rate of leaves compared with that of fruits, most of the xylemic water and Ca flow towards the leaves and away from fruit. Increasing abscisic acid (ABA) levels in the leaves may favor xylemic water and calcium uptake into the low transpiring fruit tissue and help reduce calcium disorders. Therefore, several trials have been conducted since 2012 to investigate how exogenous ABA applications would help improve calcium uptake and prevent bitter pit in apple fruit. 1) In 2012, foliar applications of 350 $mg{\cdot}L^{-1}$ S-ABA were applied to Honeycrisp apples in Wooster OH, US. 2) In 2015, foliar applications of 400 $mg{\cdot}L^{-1}$ S-ABA were applied to Honeycrisp apples in Geneva NY, US. 3) In 2016, foliar applications of 200 and 400 $mg{\cdot}L^{-1}$ S-ABA were applied to Honeycrisp in 2 different orchards in NY, including a commercial orchard. Fruit were harvested at maturity and evaluated for yield, quality and nutrient concentrations in order to assess the effects of S-ABA. Incidence (%) of bitter pit was evaluated on 40 apples per tree at harvest and after storage. In all years, foliar applications of S-ABA significantly reduced bitter pit incidence of apples in more than 50%. No negative effects of S-ABA on fruit production, fruit size or fruit quality were observed. Also, no thinning effect occurred at both rates used, though sprays were started 15 days after full bloom. Very little leaf yellowing was observed. Calcium content in the leaves tend to decrease and the uptake into the fruit tissue increased with S-ABA foliar spray treatments. So far, S-ABA applications have been showing great benefits on controlling bitter pit in apples.

EXOGENOUS ABA REDUCES LOSS OF COLD HARDINESS IN ECODORMANT BUDS OF VITIS VINIFERA

Alisson P. Kovaleski[1] and Jason P. Londo[2]

[1]Cornell University, School of Integrative Plant Science, Section of Horticulture

(ap874@cornell.edu)

[2]USDA-ARS Grape Genetics Research Unit

Grapevines (Vitis spp.) produce mixed lateral buds that must overwinter in order to produce a crop in the following season. In order to do so, buds acclimate to very low below freezing temperature in response to decreasing air temperature in the winter. In the spring, as temperatures rise, buds fully de-harden prior to growth resumption. Cold hardiness can be measured using differential thermal analysis, in which the freezing of intracellular water at a given temperature shows exotherm signatures that demonstrate the failure of the mechanism responsible to resist low temperatures. Previous data demonstrated that abscisic acid (ABA) biosynthesis is downregulated during deacclimation. Because the expression of these genes was negatively correlated with the cold hardiness phenotype, meaning that the reduction in ABA biosynthesis was correlated to a reduction of the cold hardiness, we hypothesized that applying exogenous ABA would prevent the loss of cold hardiness. To study this, single node cuttings of ecodormant V. vinifera 'Cabernet Franc' and 'Riesling' were collected from the field and placed in a growth chamber with temperatures ranging between 1.5 and 12.5 °C (7 °C average). The buds were placed in cups of water (control), 1 or 5mM ABA solutions, or sprayed with 10 or 20 mM ABA. Cold hardiness loss was measured using DTA twice a week for 8 weeks and phenology was assessed after 4 weeks of treatment. A second set of buds was placed in a growth chamber at 22 °C, with either 18/6h or 0/24h light/dark photoperiod. In the second experiment, a water control and 1mM ABA solution treatments were used. All treatments deacclimated similarly for the first two weeks. However, cuttings that were in ABA solutions stopped losing hardiness after 2 to three weeks at 7 °C, or 3 days at 22 °C, remaining at the same level of hardiness until placed in forcing conditions in case of 7 °C. Cuttings from the control continued to deacclimate. Although loss of cold hardiness was halted by ABA, the buds did not gain additional hardiness. It appears that ABA is involved in maintaining cold hardiness, but may not be necessarily involved in increasing hardiness in ecodormant buds.

INTRODUCTION

Perennial plants such as cultivated grapevines (*Vitis vinifera*) developed dormancy mechanisms in order to survive winter temperatures. In *Vitis* species, the growing vines produce latent buds in leaf axils, which are dormant due to inhibition by the growing shoot tip (paradormancy), preventing growth of lateral vines (Lang, 1987). Depending on environmental cues, these latent buds may "break" and resume growth, generating new growing vines. During the fall, the reduction in the photoperiod signals growth to discontinue in actively growing shoots, and along with reducing temperatures, initiate necessary changes to induce endodormancy in the buds that remained paradormant (Fennel and Hoover, 1991). Endodormant buds are unable to break and resume growth before exposure to low temperatures, when they fulfill their chill requirement (Lang, 1987). Chill accumulation can be calculated by a variety of methods, in which accumulation typically occurs as a result of exposure to temperatures ranging from 0 to 10 °C (Byrne and Bacon, 1992). This mechanism allows the buds to remain dormant even during warm spells, which could result in bud mortality if bud-break occurred. Once the chill requirement is fulfilled, buds transition into ecodormancy. During ecodormancy, the buds remain in a dormant state while environmental conditions are not conducive for growth (Lang, 1987) but become sensitive to warming temperatures.

During the dormancy period, grapevine bud cold hardiness can be assessed by using differential thermal analysis (DTA; Andrews et al., 1984). Through this method, the freezing of water inside buds is detected in two peaks: the high temperature exotherm, when extracellular water freezes, which is not lethal; and the low temperature exotherm (LTE), when the intracellular water freezes, and correlates with bud death (Andrews et al., 1984; Mills et al., 2006).

Cold hardiness is a dynamic trait in grapevine buds, and its variation is largely believed to be due to air temperature (Ferguson et al., 2011). During the early winter, buds go through a period of acclimation, followed by a generally stable period, and a deacclimation period in the early spring (Londo and Kovaleski, 2017). Endodormant grapevine buds are more responsive towards acclimation, gaining hardiness at temperatures below 11-15 °C, depending on the cultivar (Ferguson et al., 2011; Ferguson et al., 2013). However, ecodormant buds tend to deacclimate when temperatures are above 3-7 °C. Winter temperature variability is expected to increase with climate change (Vasseur et al., 2014), with occurrences of unseasonable warm events being more frequent. Transient warm temperatures can pose special risk to ecodormant buds, which may rapidly lose their cold hardiness and resume growth due to warmth and suffer damage due to the low temperatures following the warm spell. A previous study using RNA-Seq in deacclimating ecodormant buds of V. vinifera 'Cabernet Sauvignon' demonstrated that abscisic acid (ABA) biosynthesis is downregulated during deacclimation, with special downregulation of ABA1 and NCED3 enzymes (data not shown). Because the expression of these genes was negatively correlated with the cold hardiness phenotype, we hypothesized that applying exogenous ABA would prevent the loss of cold hardiness. Therefore, the objective of this study was to test the application of exogenous ABA to prevent loss of cold hardiness in grapevine buds.

MATERIALS AND METHODS

Single node cuttings of ecodormant V. vinifera 'Cabernet Franc' and 'Riesling' were collected from the field. The cuttings were placed in cups of water and separated into two trials.

Trial 1. Cuttings were placed in a growth chamber with temperatures ranging between 1.5 and 12.5 °C (7 °C average) and 0/24h light/dark photoperiod. ABA was applied as either a solution, in which the single node cuttings were sitting, or as a spray directly on the bud. The buds were sprayed to run-off 5 times in 5 consecutive days. The concentrations used were 1mM and 5 mM for the solution, and 10 mM and 20 mM for the sprays. After 42 days, buds were washed in running water, and placed in cups of water under 22 °C and 16/8h light/dark photoperiod. Cold hardiness loss was measured using DTA twice a week for 8 weeks or until it was no longer detectable. Phenological stage was evaluated using the modified E-L scale (Coombe, 1995) until buds reached stage 12 (5 leaves separated).

Trial 2. Cuttings were placed in a growth chamber at 22 °C. Two photoperiods were used: 16/8h or 0/24h light/dark photoperiod. ABA was applied in a solution with 1mM, and compared to a water control. Cold hardiness was measured through DTA approximately every 2 days for 10 days.

Data were compared using standard deviations from the mean, and figures were prepared using the package ggplot2 on R (ver. 3.3.0, R Foundation for Statistical Computing, Vienna, Austria).

RESULTS

Trial 1. All treatments deacclimated similarly for approximately two ('Cabernet Franc' – Figure 1, 'Riesling' – data not shown) or three weeks ('Cabernet Franc' – Figure 2, 'Riesling' – data not shown) for solution and spray applications, respectively. However, after this period, loss of cold hardiness was halted in all ABA treatments, but continued in the water control treatment. After buds had been moved to 22 °C, a quick loss of cold hardiness was observed. Buds in the water control also started growing before any of the ABA treatments, reaching budbreak (E-L stage 4) at 48 days, while it took 48 days for 1mM, and >56 days for 10 and 20 mM. The 5 mM treatment never resumed growth, possibly due to observed xylem damage to the cane.

Trial 2. All treatments deacclimated similarly until day 3 ('Riesling' – Figure 3, 'Cabernet Franc' – data not shown). After this period, buds given ABA treatments stopped losing cold hardiness, and maintained a lower LTE compared to the water control regardless of photoperiod. However, the difference between 1mM ABA and control in 16/8h was greater than in 0/24h.

DISCUSSION

ABA is known to be a stress induced plant hormone and closely linked to cold or freezing stress (Thomashow, 1999, and references within). In grapevines, exogenous application has been previously reported to enhance cold hardiness (Zhang and Dami, 2012; Dami et al., 2015; Karimi and Ershadi, 2015). However, these previous reports used foliar applications during the summer and evaluated cold damage on leaves during the fall (Karimi and Ershadi, 2015), or buds during the winter (Zhang and Dami, 2012; Dami et al., 2015). Our study looked at direct application of ABA to grapevine buds in order to enhance cold hardiness.

Cold hardiness loss was halted by ABA treatments in both temperature treatments, with applications by solution or spray. Considering that there were little to no differences between the ABA treatments, it is likely that the concentration applied can be reduced. In leaf applications of field experiments, previous studies used concentrations ranging from 0.05 to 2.3 mM (Zhang and Dami, 2012; Dami et al., 2015; Karimi and Ershadi, 2015). However, it is important to consider that the uptake by sprays in dormant tissue is likely lower or happens at a reduced rate, and that ABA can be degraded by UV light when applied in the field.

Solution application of ABA at 5 mM concentration resulted in damage of to the cane that prevented budbreak. Although this result may be important for growth chamber experiments, it is not practical to irrigate plants with ABA in the field during the winter. In the sprays, the higher concentration did not damage buds, and development occurred normally once it started.

While deacclimation is known to be temperature dependent (Ferguson et al., 2011; Kovaleski and Londo, 2016), a temperature dependence in the time to response to the ABA treatments is also clear. Within the 10 days of evaluation in 22 °C we were able to detect differences between the ABA treatments and the controls, whereas at 7 °C no differences were detected within the first 2-3 weeks. Further testing should use more temperatures in order to calculate a time to response in growing degree-days, so that these results may be used to predict ABA effect in variable temperature conditions.

It appears that ABA applications to dormant tissues of grapevines can be used to prevent loss of cold hardiness, but did not lead to a recovery to initial hardiness levels. However, further testing and field experiments are required to find the concentrations necessary, as well as the time prior to warm temperature exposure to find the results necessary to prevent cold damage following warm spells.

REFERENCES

Andrews, P.K., C.R. Sandidge III, and T.K. Toyama. 1984. Deep supercooling of dormant and deacclimating Vitis buds. Am. J. Enol. Vitic. 35:175-177.

Byrne, D.H. and T.A. Bacon. 1992. Chilling estimation; its importance and estimation. Texas Horticulturist 18:5, 8-9.

Coombe, B.G. 1995. Growth stages of the grapevine. Aust. J. Grape Wine Res. 1:100-110.

Dami, I.E., S. Li, P.A. Bowen, C.P. Bogdanoff, K.C. Shellie, and J. Willwerth. 2015. Foliar applied abscisic acid increases 'Chardonnay' grapevine bud freezing tolerance during autumn cold acclimation. HortTechnology 25:293-305.

Fennel A, and Hoover E. 1991. Photoperiod influences growth, bud dormancy, and cold acclimation in Vitis labruscana and V. riparia. J. Amer. Soc. Hort. Sci. 116:270-273.

Ferguson, J.C., J.M. Tarara, L.J. Mills, G.G. Grove and M. Keller. 2011. Dynamic thermal time model of cold hardiness for dormant grapevine buds. Ann. Bot. 7:389-396.

Ferguson, J.C., M.M. Moyer, L.J. Mills, G. Hoogenboom and M. Keller. 2013. Modeling dormant bud cold hardiness and budbreak in 23 Vitis genotypes reveals variation by region of origin. Am. J. Enol. Vitic. Doi:10.5344/ajev.2013.13098

Karimi, R. and A. Ershadi. 2015. Role of exogenous abscisic acid in adapting of 'Sultana' grapevine to low-temperature stress. Acta Physiol. Plant. 37:151.

Kovaleski, A.P., J.P. Londo. 2016. Could base temperatures be negative for grapes? Deacclimation of cultivated and wild grapes in controlled environment. ASHS Annual Conference. HortScience 51:S151.

Lang, G.A. 1987. Dormancy: a new universal terminology. HortScience 22:817-820.

Londo, J.P. and A.P. Kovaleski. 2017. Characterization of wild North American grapevine cold hardiness using differential thermal analysis. Am. J. Enol. Vitic. 68:203-212

Mills L.J., J.C. Ferguson and M. Keller. 2006. Cold-hardiness evaluation of grapevine buds and cane tissues. Am. J. Enol. Vitic. 57:194-200.

Vasseur, D.A, J.P. DeLong, B. Gilbert, H.S. Greig, C.D.G. Harley, K.S. McCann, V. Savage, T.D. Tunney, M.I. O'Conner. 2014. Increased temperature variation poses a greater risk to species than climate warming. Proc. R. Soc. B. DOI: 10.1098/rspb.2013.2612

Zhang, Y. and I.E. Dami. 2012. Foliar application of abscisic acid increases freezing tolerance of field-grown Vitis vinifera Cabernet franc grapevines. Am. J. Enol. Vitic. 63:377-384.

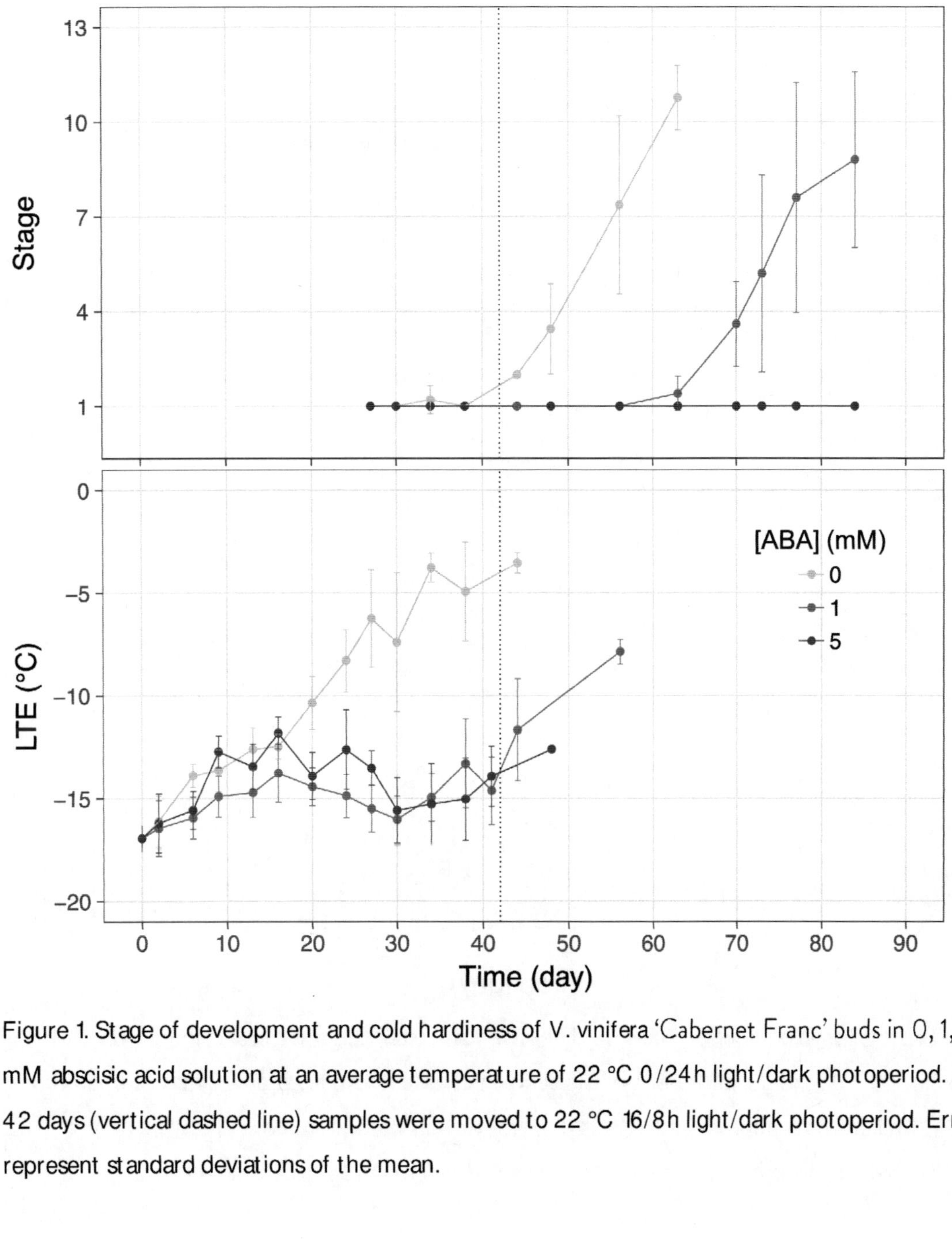

Figure 1. Stage of development and cold hardiness of V. vinifera 'Cabernet Franc' buds in 0, 1, and 5 mM abscisic acid solution at an average temperature of 22 °C 0/24h light/dark photoperiod. After 42 days (vertical dashed line) samples were moved to 22 °C 16/8h light/dark photoperiod. Error bars represent standard deviations of the mean.

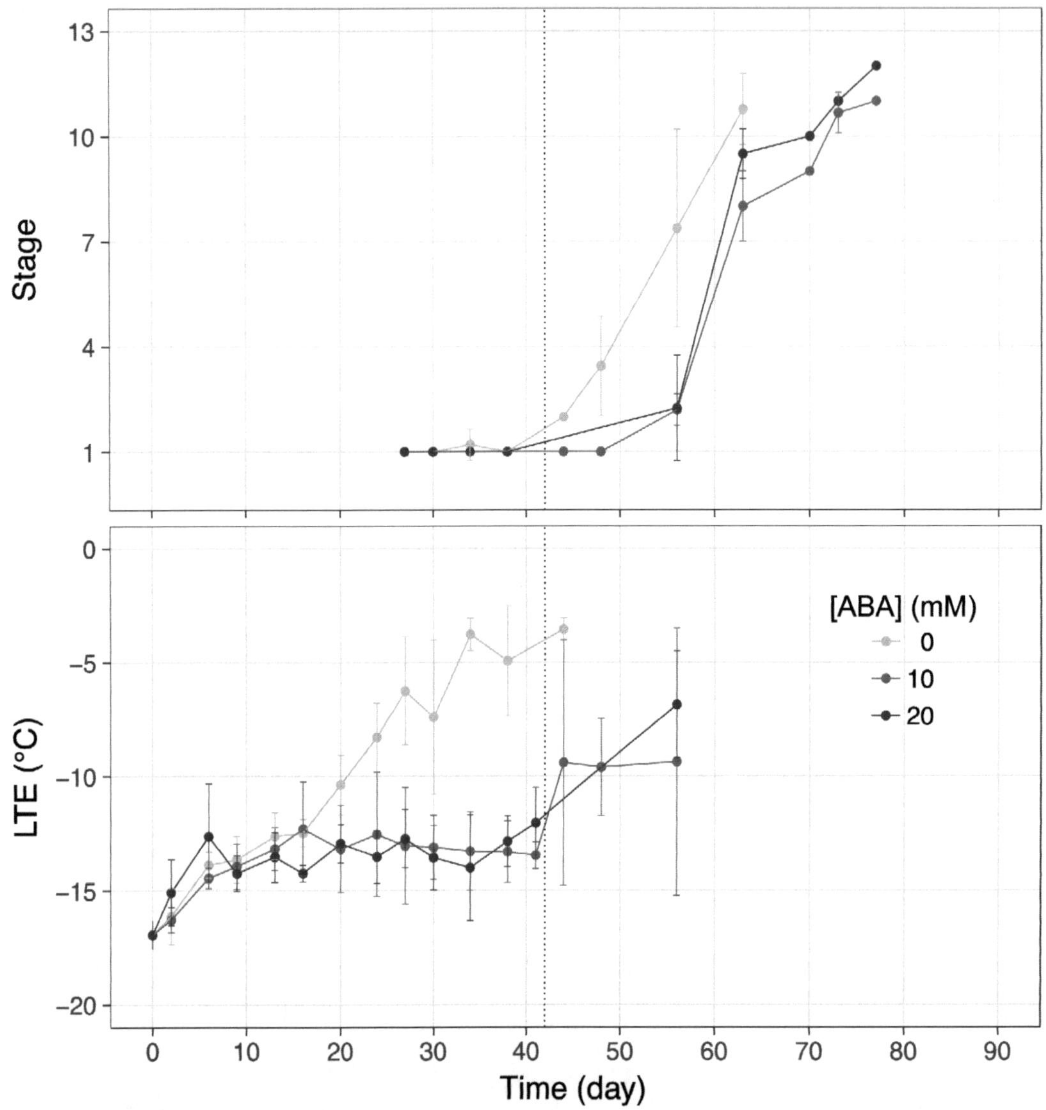

Figure 2. Stage of development and cold hardiness of V. vinifera 'Cabernet Franc' buds after 5 applications of 0, 10, and 20 mM abscisic acid spray in consecutive days at an average temperature of 7 °C 0/24h light/dark photoperiod. After 42 days (vertical dashed line) samples were moved to 22 °C 16/8h light/dark photoperiod. Error bars represent standard deviations of the mean.

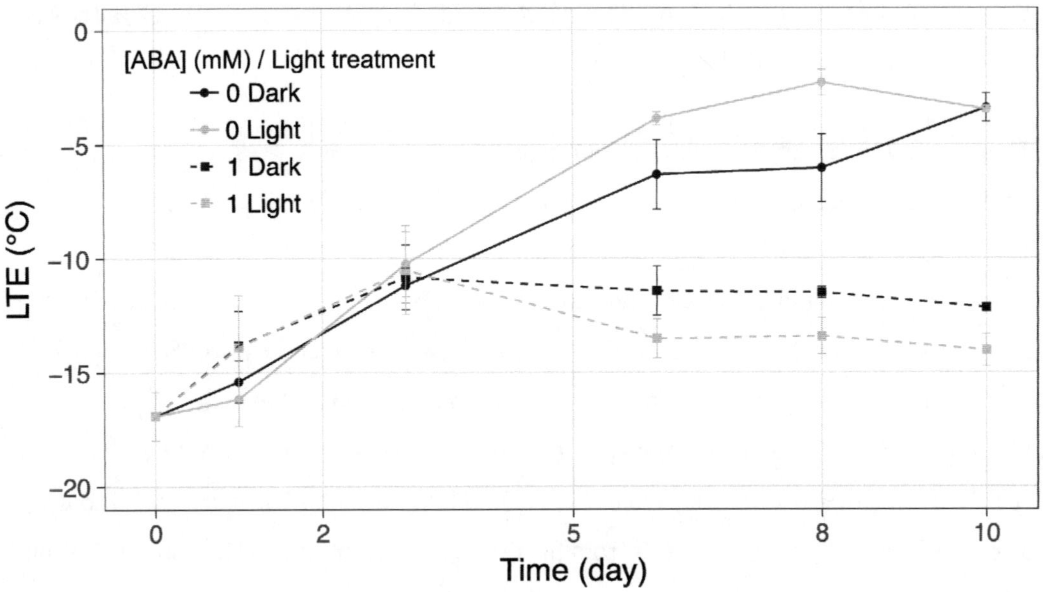

Figure 3. Cold hardiness of V. vinifera 'Riesling' buds in 0 or 1 mM abscisic acid solution at 22 °C at either 16/8 h or 0/24 h light/dark photoperiod. Error bars represent standard deviations of the mean.

FOLIAR APPLICATION OF ABSCISIC ACID TO DAY-NEUTRAL STRAWBERRY CULTIVARS TO INFLUENCE FRUIT PRODUCTION

Dennis E. Deyton, Carl E. Sams, and John C. Cummins

Department of Plant Sciences, The University of Tennessee, Knoxville, TN 37996, USA

Research was conducted on strawberry plants grown in pots in a greenhouse to determine if repeated sprays of abscisic acid (ABA) effected vegetative growth and fruit quality. 'Seascape' and 'San Andreas' day-neutral frigo strawberry plants were planted on 9 June 2016 in 3.8 L pots containing 1:1 mixture of perlite and peat moss. Plants were sprayed after first anthesis with 0, 5, 10, 15, 25, 50 or 100 mg·L^{-1} of ABA biweekly from 25 August 2016 through 22 March 2017. Plants were harvested approximately twice per week from 30 September 2016 until 31 March 2017. Fruit were graded into commercial (>10 g) and non-commercial (<10 g and deformed fruit) and weighed and counted at each harvest. ABA sprays of 25 to 100 mg·L^{-1} reduced the total number and weight of 'Seascape' and 'San Andreas' fruit harvested. Commercial fruit yield (g/plant) of 'San Andreas' plants treated with 50 or 100 mg·L^{-1} was reduced 18% and 50%, respectively, compared to untreated plants. 'Seascape' plants treated with 50 or 100 mg·L^{-1} had 40% and 70%, respectively, less commercial yield than control plants. ABA did not affect fruit soluble solid content and had little effect on berry size. Slight phytotoxicity of older leaves was noted on plants treated with >10 ABA. Treatments did not affect apparent blossom mortality, but plants sprayed with 25, 50, or 100 mg·L^{-1} ABA had 21%, 23%, and 48% fewer number of blossoms than untreated plants.

INTRODUCTION

Foliar sprays of abscisic acid (ABA) were reported to increase red color development (Cantin et al., 2007) and anthocyanin content (Koyama et al., 2014; Peppi et al., 2006) of grapes and allow earlier harvest. Weekly sprays of ABA from anthesis until harvest increased soluble sugar content in tomato fruit (Barickman et al., 2017). Chen et al. (2011) found that strawberry fruit dipped in ABA solution had enhanced FaASR gene expression and concluded that the treatment might contribute to acceleration of strawberry fruit ripening. Ayub et al. (2016) injected ABA into field-grown strawberry fruit at green, white, pink and red stages of development to determine the effects on gene expression and fruit quality. ABA increased soluble solids content in all stages after 24 h.

The objectives of this project were to determine if repeated ABA sprays throughout a production season influenced strawberry plant growth, yield and fruit soluble solid content.

MATERIALS AND METHODS

'San Andreas' and 'Seascape' day-neutral frigo strawberry plants were planted on 9 June 2016 in 3.8 L pots containing 1:1 mixture of perlite and peat moss (Fig. 1). Plants were grown in a greenhouse with fertigation supplying 80 mg/L of nitrogen (in a modified half strength Hoagland solution) after the third week of establishment. They were sprayed with 0, 5, 10, 15, 25, 50 or 100 mg·L^{-1} of s-abscisic acid (ABA) biweekly from 25 August 2016 through 22 March 2017. Treatments were randomized in a randomized complete block design with split plot array of treatments in four replications of two cultivars (main plot), seven ABA treatments (sub plot) and four plants per replication. Plants within each treatment were removed from the trial and sprayed until first runoff with a small hand-held sprayer (Fig. 2). The media surface was shielded with paper towels to reduce root contact with ABA. Paper towels covering the pot media were replaced after each spray application.

High rates of ABA in a previous trial caused phytotoxicity (yellowing) of basal leaves, thus leaf phytotoxicity was rated on 7, 15 and 26 September 2016 using the scale: 1 = no damage, 2 = slight damage, 3 = slight to medium damage, 4 = medium damage, 5 = severe damage. Daily temperatures in the greenhouse averaged 27°C from first spraying until rating. We removed leaves showing yellowing before the next spraying. Runners were counted and removed weekly from 30 August 2016 until 7 February 2017. We observed damage on blossoms and counted damaged and non-damaged flowers on four plants/plot on 3, 10, and 29 March and removed damaged flowers after each count. Plants were harvested approximately twice per week from 30 September 2016 until 31 March 2017. Fruit were graded into commercial (>10 g) and non-commercial (<10 g and deformed fruit) grades and weighed and counted at each harvest. Red fruit were sampled from each plot in January and late March and soluble solid content of composite samples (>45 g) of each time period was measured with a hand-held refractometer (Model RHB-18/ATC, Spectrum Technologies, Inc., Plainfield IL, USA).

RESULTS

VEGETATIVE PHYTOTOXICITY: Untreated strawberry plants had none or slight development of yellowing of basal leaves during the season. (Fig. 3). However, foliar phytotoxicity as displayed as development yellow basal leaves increased as ABA concentration increased from 5 to 100 mg·L^{-1} (Table 1). Newer leaves never displayed the yellowing. The yellow leaves were removed before the next spray application for two reasons. First, removal of yellow leaves allowed better evaluation of phytotoxicity after the next treatment. We do not know if the leaves would have died or may have recovered their health. We suspected that they would further decline, be more susceptible pathogen infection and compromise the results. The removable of more yellow leaves from ABA treated plants may have affected the eventual plant partition dry weights (data not shown) and may have partly influenced the reduced fruit yield of ABA treated plants (Table 2). The ABA treatments did not affect number of runners produced (Table 1).

BLOSSOM DAMAGE: We observed browning of blossoms and. The discoloration may have been due to phytotoxicity or normal decline of blossom vigor. We suspected that the yield reduction caused by ABA might partially be due to damage to flowers. We visually rated the discoloration on a 1 to 5 scale on 2 March 2017. Our visual ratings indicated more flower discoloration (we called damaged) on ABA treated plants than on control plants (data not shown). We counted the numbers of damaged and undamaged blossoms on 3, 10, and 29 March. Surprising to us, plants sprayed with 50 or 100 $mg \cdot L^{-1}$ ABA had fewer damaged blossoms/plant than control plants (Table 1). However, plants sprayed with 25, 50 or 100 $mg \cdot L^{-1}$ ABA had 21%, 23% and 48%, respectively, fewer total blossoms/plant than untreated plants. Sprays of 5 to 100 $mg \cdot L^{-1}$ ABA did not significantly affect percentage of damaged blossoms. Thus, higher rates of ABA did not increase damage on blossoms compared to the control but reduced the number of flowers that developed and reduced potential yield.

FRUIT YIELD AND QUALITY: Yields shown in Table 2 are from 30 September 2016 until 31 March 2017. Fruit yields during October, November, and December were lower than we had previously seen from August or September planted plug plants grown in the greenhouse as an off-season crop. The hydroponic nutrient supplied to the frigo plants in this trial was apparently too large, the plants became too vigorous, and fruiting was delayed. Nutrient supply was then reduced and most of the fruit yield in Table 2 occurred from late December until 31 March.

'San Andreas' plants generally produced larger fruit and more total fruit than 'Seascape' plants (Table 2). The interaction effects of cultivar and ABA on yield, berry size, and soluble solid content was frequently non-significant, and thus mean effects across cultivar are shown in the tables. Evaluation of total fruit yield by mean comparison (Fisher's protected least significant difference) showed that plants sprayed biweekly with 15 to 100 $mg \cdot L^{-1}$ ABA yielded fewer fruit and less total fruit weight than untreated plants. Commercial fruit yields (> 10 g) was less on plants sprayed with

50 or 100 mg·L^{-1} ABA than on control plants. Yields (g/plant and no/plant) of commercial and small size fruit decreased linearly as ABA concentration increased from 0 to 100 mg·L^{-1} ($P \leq 0.001$).

Commercial fruit yield (g/plant) of 'San Andreas' plants treated with 50 or 100 mg·L^{-1} was reduced 18% and 50%, respectively, compared to untreated plants (data not shown). Commercial yield of 'Seascape' plants treated with 50 or 100 mg·L^{-1} was 40% and 70%, respectively, less than control plants. 'Seascape' had smaller fruit than 'San Andreas' plants; and, as size was reduced, a larger percentage of fruit were graded into the < 10 g group (data not shown).

There was a trend for fruit on plants treated with ABA to be larger than fruit on untreated plants, but the differences were not significant in this data set. ABA did not significantly affect fruit soluble solid content and had little effect on berry size. We did not measure color but saw no visible effect of ABA on red color development on fruit.

CONCLUSIONS

Repeated sprays of > 25 mg·L^{-1} ABA on greenhouse grown strawberry plants caused yellowing of basal leaves, and reduced the number of flowers/plant and fruit/plant without significant effect on fruit soluble solid content or fruit size.

We later collected more data on blossom, runner and lateral crown numbers; yield, and in early June, partitioned the plants for fresh and dry weights of leaves, crowns, flowers and fruit. The remaining data had not been analyzed in time for this report.

LITERATURE CITED

Ayuba, R.A., L. Bosetto, C.W. Galvão, R.M. Etto, J. Inaba, P.Z. Lopes. 2016. Abscisic acid involvement on expression of related gene and phytochemicals during ripening in strawberry fruit Fragaria x ananassa cv. Camino Real. Scientia Horticulturae 203:178–184. https://doi.org/10.1016/j.scienta.2016.03.026.

Barickman, T.C., D.A. Kopsell and C.E. Sams. 2017. Abscisic acid improves tomato fruit quality by increasing soluble sugar concentrations. Journal of Plant Nutrition 40:964-973.

Cantin, C.M., M.W. Fidelibus, and C.H. Crisostoc. 2007. Application of abscisic acid (ABA) at veraison advanced red color development and maintained postharvest quality of 'Crimson Seedless' grapes. Postharvest Biology and Technology 46:237–241.

Chen, J., D. Liu, Y. Jiang, M. Zhao, W. Shan, J. Kuang, and W. Lu. 2011. Molecular characterization of a strawberry FaASR gene in relation to fruit ripening. PLoS One 6, e24649, http://dx.doi.org/10.1371/journal.pone.0024649.

Koyama, R., A.M. de Assis, L.Y. Yamamoto, W.F. Borges and R. de Sá Borges. 2014. Exogenous abscisic acid increases the anthocyanin concentration of berry and juice from 'Isabel' grapes (Vitis labrusca L.). HortScience 49:460-464.

Peppi, M.C., M.W. Fidelibus, and N. Dokoozlian. 2006. Abscisic acid application timing and concentration affect firmness, pigmentation and color of 'Flame Seedless' grapes. HortScience 41:1440 – 1445.

Table 1. Phytotoxicity, runners, and flowering of 'San Andreas' and 'Seascape' plants sprayed biweekly with ABA

	Leaf phytotoxicity[Z]	Runners (no./plant)[Y]	Blossoms		
			Damaged[W] (no./plant)	Total (no./plant)	Damaged (%)[V]
Cultivar					
'San Andreas'	2.0 a[V]	12.8 a	11.3 a	39.9 a	27.8 a
'Seascape'	2.2 b	5.4 b	7.1 b	35.1 b	20.5 b
ABA ($mg \cdot L^{-1}$)					
0	1.3 e	7.7 a	12.0 a	45.8 a	26.2 ab
5	1.9 c	9.9 a	9.3 ab	42.7 ab	21.5 a
10	2.1 c	8.5 a	10.9 a	38.8 bc	28.2 ab
15	2.2 cd	9.9 a	10.1 ab	40.5 abc	25.6 ab
25	2.2 bc	8.3 a	9.5 ab	36.0 bc	25.5 ab
50	2.5 b	10.5 a	6.8 bc	35.2 c	18.9 b
100	3.2 a	8.9 a	5.6 c	23.6 d	23.2 ab
P-value[U]					
Cultivar (C)	***	***	***	**	***
ABA	***	NS	**	***	NS
Linear	NS	NS	NS	NS	NS
Quadratic	NS	NS	NS	NS	NS
C x ABA	NS	NS	NS	NS	NS

[Z]Leaf phytotoxicity, average of 4 plants/plot rated on 7, 15 and 26 September 2016. Scale: 1=no damage, 2=slight damage, 3=slight to medium damage, 4 = medium damage, 5=severe damage.

[Y]Runners removed from 30 August 2016 until 7 February 2017.

[W]Means of damaged and total number of blossoms counted on 3, 10, and 29 March 2017.

[V]Means within a column and category followed by the same letter are not significantly different according to Fisher's protected least significant difference ($P \leq 0.05$).

NS, *, **, *** Non-significant or significant at $P \leq 0.05$, 0.01 or 0.001, respectively.

Table 2. Fruit yield and quality of 'San Andreas' and 'Seascape' strawberry plants sprayed biweekly with ABA

	Yield (g/plant)^Z			Berries (no. /plant)			Fruit size	SS^Y
	$Commercial^X$	$Small^W$	Total	Commercial	Small	Total	(g/berry)	(%)
Cultivar (C)								
'San Andreas'	339.1 a	104.5 a	443.6 a	20.5 a	17.9 a	38.5 a	11.7 a	7.91 a
'Seascape'	196.9 a	201.7 a	398.6 b	14.1 b	37.3 b	51.5 b	7.8 b	8.43 b
ABA ($mg \cdot L^{-1}$)^V								
0	318.7 a^V	211.6 a	530.3 a	21.0 a	39.7 a	59.7 a	9.1 b	8.39 a
5	308.7 a	174.2 a	482.9 ab	20.3 ab	31.5 b	51.4 ab	9.6 ab	8.09 a
10	324.1 a	200.8 a	524.9 ab	20.7 a	36.4 ab	57.1 a	9.7 ab	8.55 a
15	281.1 a	179.8 a	460.8 bc	18.3 ab	31.2 b	49.7 b	9.7 ab	8.05 a
25	279.0 ab	121.5 b	400.5 cd	17.8 b	21.3 c	40.3 c	10.3 a	8.20 a
50	231.9 b	129.1 b	361.0 d	14.6 c	23.2 c	38.8 c	9.8 ab	7.96 a
100	132.6 c	54.7 c	187.3 e	8.2 d	10.0 d	18.4 d	10.1 ab	7.94 a
P-value^U								
Cultivar (C)	***	***	**	***	***	***	***	*
ABA	***	***	***	***	***	***	NS	NS
Linear	***	***	***	***	***	***	NS	NS
Quadratic	NS	NS	NS	NS	NS	NS	NS	NS
C x ABA	NS	*	NS	NS	*	*	NS	NS

^{Z}Yield from 30 September 2016 until 31 March 2017.

^{Y}Means of Soluble Solids Content measured on 23 January and 31 March 2017, composite samples > 45 g.

^{X}Fruit larger than 10 g and of commercial quality.

^{W}Fruit smaller than 10 g, and a few cull fruit.

^{V}Foliar sprays applied biweekly from 25 August 2016 until 22 March 2017.

^{U}Means within a column and category followed by the same letter are not significantly different according to Fisher's protected least significant difference ($P \leq 0.05$). NS, *, **, *** Non-significant or significant at $P \leq 0.05$, 0.01 or 0.001, respectively.

Figure 1. Established frigo strawberry plants for treatment with ABA concentrations after first anthesis.

Figure 2. Plants of each replicate were removed from the trial and hand sprayed with the appropriate ABA treatment.

Figure 1. Yellowing of basal leaves of plant sprayed with ABA.

Figure 2. Appearance of plants on (A) 3 October 2016 and (B) on 11 January 2017.

IN VITRO SHOOT AND ROOT REGENERATION OF PEARLBUSH UNDER PGR TREATMENTS

Guochen Yang, Zhongge (Cindy) Lu, and Julia Robinson
Department of Natural Resources & Environmental Design, North Carolina A&T State University, Greensboro, NC 27411.
E-mail: yangg@ncat.edu

Pearlbush (Exochorda racemosa) is an ornamental shrub with white flowers that is spiraea-like, deciduous and hardy, and is difficult to propagate. Normally it is propagated by seeds, layering, or softwood cuttings. However, these methods generally take several years to produce marketable plants. Because of the relatively long developmental period, methods that can accelerate the growth period and increase the number of plantlets could greatly benefit the plants' commercial value. Micropropagation is a proven and more efficient method for propagating many woody plant species, therefore could potentially expedite production of pearlbush. In this study, we tested different plant growth regulators on their impacts for shoot initiation and proliferation and rooting of micropropagated pearlbush plantlets. Different concentrations of kinetin, zeatin, thidiazuron (TDZ), 2,4-D, 2,4,5-T, IBA or NAA were added to WPM basal medium supplemented with 3% sucrose and 0.6% agar, pH 5.8. All cultures were transferred onto fresh media every 4 weeks and incubated under a 16-hour photoperiod and a photo flux density of 12.8 ± 4.8 Mol $s^{-1}m^{-2}$ light provided by cool white fluorescent tube at 23°C. Our results indicated that explants cultured on WPM with inclusion of kinetin or TDZ produced the best shoot initiation and proliferation and best rooting with inclusion of 2,4-D or NAA. The plantlets were rooted in vitro and subsequently transferred to peat-lite substrate and acclimatized in the plant tissue culture laboratory. Acclimatized plantlets were then moved to a greenhouse on campus and allowed to grow under normal conditions. The plants grow vigorously with very healthy morphology.

INTRODUCTION

Pearlbush (Exochorda racemosa) is an ornamental shrub and native to China and produces plentiful spiraea-like white flowers (Figure 1). It is a loose, upright, early-blooming, and deciduous shrub with a broad rounded shape canopy, and could grow up to 10 - 15 feet tall. The flower buds resemble white pearls, therefore the common name. Flowers bloom in spring as early as in February in piedmont triad area of North Carolina. The leaves are light green, narrow with elliptic to oblong shape and is about 3" in length. It is an excellent choice for early spring landscaping, especially for borders, as it is floriferous and attractive in very early spring while other plants are still dormant. It is very hardy and heat and drought tolerant once established (Plant Hardiness zones 4-8), and can be severely pruned for rejuvenating preferably after blooming is done. Spongberg (1993) reported that pearlbush is among the list of Fortune plant introductions from Eastern Asia. Pearlbush is also among the populous edible wild herbs due to its high nutrient values in its young tender leaves and buds (Pei and Du 2003). It has been confirmed that the nutritional composition of pearlbush plants has twenty confirmed phytochemicals (Zhang et al 2011). The good nutritional value of pearlbush plants is also due to the higher contents of linoleic acid (19.56%) in the seed oil (Matthaus and Ozcan 2014). Because of its beneficial properties on the skin, linoleic acid has become increasingly popular in the cosmetic industry. Research findings confirmed that linoleic acid also has anti-inflammatory, acne reductive, and moisture retentive properties when applied topically on the skin (Diezel et al 1993, Letawe et al 1998, Darmstadt et al 2002). Also, extracts of pearlbush plants demonstrated antibacterial activity (Jantova et al 2000). Normally, pearlbush is propagated by seeds, layers, or softwood cuttings. However, it is a slow process that takes several years to produce marketable plants. Therefore, this project was initiated with a goal to develop an efficient micropropagation protocol to successfully propagate this recalcitrant ornamental species.

MATERIALS AND METHODS

Softwood explant materials were collected from a local nursery and disinfested with 15% bleach solution (0.78% NaOCl) for 15 minutes and rinsed three times with sterile distilled, deionized water. In vitro cultures were established and maintained in Woody Plant Medium (WPM, Lloyd and McCown 1981) supplemented with 0.1 mg/l BA, 3% sucrose and 0.6% agar at pH 5.8.

Softwood shoot segments (0.5 cm in length) from the stock cultures were transferred to each test tube containing 10 ml of WPM supplemented with kinetin, thidiazuron (TDZ) or zeatin at 0.1, 0.5, 1.0, and 2.0 $mg{\cdot}L^{-1}$ to evaluate their effects on shoot initiation and proliferation, or with 2,4-D, 2,4,5-D, IBA or NAA at 0.5, 1.0, 1.5 and 2.0 $mg{\cdot}L^{-1}$ to evaluate their impacts on rooting of micropropagated pearlbush plantlets. Cultures were placed in the culture shelves with 16-hr photoperiod provided by cool white fluorescent tubes at 23± 1°C, and transferred to fresh media every 4 weeks.

RESULTS AND DISCUSSION

Shoot initiation was achieved from all cultures containing different concentrations of kinetin, TDZ, or zeatin (Figure 2). However, differences were observed on plantlets in terms of number of shoots per explant among different plant growth regulators (PGR) and their concentrations within a particular PGR. Explants cultured in medium containing kinetin at all concentrations tested (0.1, 0.5, 1.5, and 2.0 mg·L^{-1}) produced multiple shoots, with the best shoot proliferation observed from explants cultured in medium containing kinetin at 1.5 or 2.0 mg/l. Multiple shoot proliferation was observed from explants cultured in medium containing TDZ at a lower concentration (0.1 mg·L^{-1}). Shoot growth was observed from explants cultured in medium containing zeatin, with no observed shoot multiplication.

In vitro rooting of pearlbush was also observed from all cultures containing different concentrations of NAA, 2,4-D, or 2,4,5-T (Figure 3). However, significant differences on rooting (morphology, health and number) was noticed among plant growth regulators and concentrations within a particular PGR. From the testing IBA on stimulating root initiation, no rooting was observed. Instead, good shoot growth was observed from explants cultured in medium containing IBA at 0.5, 1.0, and 1.5 mg·L^{-1}. Also, explants cultured in the medium containing IBA at 1.5 mg·L^{-1} produced multiple shoots. This might be due to the fact that stock cultures were maintained in WPM containing a low concentration of BA. Therefore, the interaction of added IBA in the rooting medium with the BA in the explants primed from stock cultures become favorable for shoot growth and proliferation. As often the case, certain ratio of auxin and cytokinin is required for shoot or root initiation depending on plant species. More research is needed here. Healthy rooting was observed from explants cultured in medium containing NAA at 0.5 mg·L^{-1}. Rooting was also observed from treatments of NAA at 1.0 - 2.0 mg·L^{-1}, but rooting was not as healthy as those observed from NAA at 0.5 mg·L^{-1}. Reasonable rooting was also observed from explants cultured in medium containing 2,4-D at 0.5 and 1.0 mg·L^{-1} which are better than the concentrations at 1.5, and 2.0 mg·L^{-1}. Some rooting was observed from explants cultured in medium containing 2,4,5-T, but much less than those from treatments of NAA and 2,4-D. Among all auxins tested, NAA at 0.5 mg·L^{-1} was the best in stimulating healthy rooting of pearlbush.

While using a different culture medium (MS medium), Zhou et al (2008) tested BA and kinetin and NAA and IBA for induction of rapid bud propagation and reported the combination of lower concentrations of IBA at 0.3 mg·L^{-1} and BA at 0.5 mg·L^{-1} produced the best bud propagation and half strength MS plus NAA at 0.3 mg·L^{-1} stimulated rooting of explants. This is similar to our observations on shoot growth from our experiments with IBA as our stock cultures were maintained with a low concentration of BA. Similar shoot growth was observed using IBA from two different experimental procedures and culture media. According to Zhou et al (2008) and our data, NAA at a low concentration (around 0.5 mg·L^{-1}) would be the best for in vitro rooting of pearlbush. Our pearlbush plantlets produced from in vitro culture were acclimatized in the plant biotechnology laboratory (Figure 4) and subsequently growing vigorously with normal morphology under normal greenhouse conditions (Figure 5).

REFERENCES

Darmstadt GL, M Mao-Qiang, E Chi, SK Saha, VA Ziboh, RE Black, M Santosham, and PM Elias 2002 "Impact of topical oils on the skin barrier: possible implications for neonatal health in developing countries." Acta Paediatrica. 91(5): 546–554. PMID 12113324. doi:10.1080/080352502753711678.

Diezel WE, E Schulz, M Skanks, and H Heise 1993 "Plant oils: Topical application and anti-inflammatory effects (croton oil test)." Dermatologische Monatsschrift. 179: 173.

Jantova S, M Nagy, L Ruzekova, and D Grancai 2000 Antibacterial activity of plant extracts from the families Fabaceae, Oleaceae, Philadelphaceae, Rosaceae and Staphyleaceae. December 2000, 14(8): 601-603.

Letawe C, M Boone, and GE Pierard, GE 1998 "Digital image analysis of the effect of topically applied linoleic acid on acne microcomedones." Clinical & Experimental Dermatology. 23 (2): 56–58. PMID 9692305. doi:10.1046/j.1365-2230.1998.00315.x.

Lloyd G & BH McCown 1981 Commercially feasible micropropagation of mountain laurel (Kalmia latifolia) by use of shoot tip culture. Proc. Int. Plant Prop. Soc. 30:421-427.

Matthaus B and MM Ozcan 2014 Fatty acid, tocopherol and squalene contents of Rosaceae seed oils. Botanical Studies 55:48. DOI:10.1186/s40529-014-0048-4.

Pei H and X Du 2003 Analysis on the nutrients of mountainous edible wild herb Exochorda racemose. Chinese Wild Plant Resources 2003-01.

Spongberg SA 1993 Exploration and introduction of ornamental and landscape plants from eastern Asia. P140-147. In J Janick and JE Simons (eds.), New Crops. Wiley, New York.

Zhang J, X Li, L Ren, C Fang, and F Wang 2011 Chemical Constituents from Exochorda racemosa. Journal of Chinese Medicine. May 2011, 36(9): 1198-201.

Zhou L et al 2008 Study on rapid propagation technique of Exochorda racemosa in vitro. Journal of Anhui Ag Sci. 2008-19.

Figures 1-5 (L-R): Pearlbush plants in bloom; Shoot clusters from tissue culture; In vitro rooted pearlbush plantlet; Acclimatization of micropropagated pearlbush plantlets; and Healthy and vigorously growing micropropgagated pearlbush plants in one-gallon pots.

PHYSIOLOGICAL EFFECT OF RHIZOSPHERICAL BACTERIAL STRAINS IN PRODUCTS VOODOO JUICE AND TARANTULA

Emiliyan Dzhustov[1], Ivaylo Siromahov[2], Tsvetana Licheva[1,2], Rositsa Tropcheva[1], Valentin Savov[1]

[1]Sofia University St. Kliment Ohridski, Faculty of Biology, 8 "Dragan Tsankov" str., 1164 Sofia, Bulgaria

[2]Balkan Plant Science Ltd., zh.k. Hanove 1, 2060 Vakarel, Bulgaria

The aim of the study is to determine the phytohormonal effect of products Voodoo Juice and Tarantula produced by Advanced Nutrients Ltd. These products contain combinations of rhizospheric bacterial strains known to synthesize biologically active substances (BAS) like phytohormones, which support plant growth and development; increase the productivity of photosynthesis; enhance enzymatic processes in plants and improve water regime; increase the activity of other microorganisms in soil; act as antagonists of phytopathogenic microorganisms and increase the protective ability of plants. The CFU/ml is 10 10 for Voodoo Juice and 10 9 for Tarantula.

For the purpose of the study, a chromatographic analysis was performed and the products were used in a pot test with broccoli (*Brassica oleracea*). High Performance Liquid Chromatography tandem mass spectrometry was used for the chromatographic analysis.

The results show the presence of GA and paclobutrazol for both products (Voodoo Juice and Tarantula). The high concentration of phytohormones from the gibberellin family (up to 49.22 ng/ml GA4) indicates the potential of the studied products to help germination, increase plant leaf area and stem height and influence the overall development of plants. The experiment with broccoli confirms that the treatment has a certain effect on the development of the test plants, which is specific for the individual products.

HOSTA SHOOT ENHANCEMENT WITH BENZYLADENINE SPRAYS AND DRENCHES

Josh B. Henry, Brian E. Whipker, and Ingram McCall
Department of Horticultural Science, North Carolina State University, Raleigh, NC 27606 U.S.A.

Benzyladenine (Configure; Fine Americas) is used to enhance axillary shoot development in ornamental plants. Producing full and lush plants is essential for marketability of ornamental species, especially those grown primarily for their foliage. This study investigated the effect of benzyladenine sprays and drenches on the axillary shoot development of hosta [Hosta 'June' (Tardiana Group)]. Overwintered hosta plants received either a single drench of 250, 500, 1,000, or 2,000 mg·L^{-1} benzyladenine or two spray applications of 500, 1,000, or 1,500 mg·L^{-1} benzyladenine, applied one week apart. Plants were grown out for eight weeks, and the subsequent number of shoots was then determined and compared among these various benzyladenine applications. Spray applications did not significantly increase shoot development, but all drench concentrations effectively increased shoot number. In fact, plants receiving a drench of 1,000 or 2,000 mg·L^{-1} benzyladenine developed roughly 8 – 10 × the number of axillary shoots compared to untreated plants. These results indicate that benzyladenine drenches led to increased axillary shoots, and consequently improved pot fill.

INTRODUCTION

Benzyladenine (BA; Configure; Fine Americas) is used to enhance axillary shoot development in ornamental plants. Producing full and lush plants is essential for marketability of ornamental species that are grown for their foliage. Hosta (Hosta) is one such crop that can benefit from increased axillary shoot development to improve pot fill. Previous studies investigated BA applications to improve hosta offshoot development for propagation purposes (Keever, 1994), and to improve summer quality in containers and landscapes (Schultz et al., 2000). These studies indicated that sprays and drenches were effective in enhancing axillary shoot development; however, drenches have been reported to cause significant phytotoxicity (Keever, 1994). Configure is currently labeled for hosta production, but only as a spray. This study was conducted to compare the efficacy of BA sprays and drenches on axillary shoot development of hosta.

MATERIALS AND METHODS

Overwintered Hosta 'June' (Tardiana Group) plants were grown in a polyethylene glazed greenhouse in Raleigh, NC (35° N Latitude). The plants were grown in 1 gallon nursery pots filled with a bark-based substrate. Temperature set points were 18°/13°C (day/night). Plants were automatically irrigated as needed via drip tubes, and fertilizer was applied using a 13N –0.8P–10.8K fertilizer mixed at 150 mg·L^{-1} N. Axillary shoots were counted for each plant prior to BA applications. Plants received either a single drench application of 250, 500, 1,000, or 2,000 mg·L^{-1} BA or two sprays of 500, 1,000, or 1,500 mg·L^{-1} BA applied one week apart. There were eight single plant replicates of treatment. Final shoot number was quantified again eight weeks after the initial BA application, and the study was terminated.

Data were analyzed with SAS (version 9.4; SAS Institute, Cary, NC) using PROC GLM to determine differences in axillary shoot development. Means were separated using the Tukey adjustment. Data were then subjected to PROC REG to determine the best fit linear or quadratic model based on r2.

RESULTS AND DISCUSSION

Hosta plants that did not receive a BA application did not develop any additional axillary shoots over the eight weeks of the study (Fig. 1). BA sprays did not significantly affect axillary shoot development, as axillary shoot numbers were similar to the untreated control (Fig. 1). As such, no regression model was selected for BA sprays. The data in conjunction with visual observations (Fig. 2A) indicate that sprays were ineffective in this study. These results differ from previous research which found single application sprays of 1,250–3,750 mg·L^{-1} BA effectively enhanced axillary shoot formation in 10 hosta cultivars (Garner, 1997). Single application sprays of 2,000–3,000 mg·L^{-1} BA were reported by Keever (1994) as an effective method of axillary shoot development in hosta as well. This may indicate that a higher single spray application was a superior method to two lower spray applications one week apart.

All BA drenches significantly affected axillary shoot formation, and the response was best fit with a quadratic regression model (Fig. 1). Drenches of 250 and 500 mg·L^{-1} BA resulted in similar axillary shoot development greater than the untreated control (Fig. 1). Drenches of 1,000 and 2,000 mg·L^{-1} BA developed the highest number of axillary shoots, with 8–10 × the number of shoots compared to untreated plants (Fig. 1). Additionally, the phytotoxicity symptoms reported in response to BA drenches in an earlier study (Keever, 1994) were not present on plants drenched with BA in this study (Fig. 2B). These results indicate that BA drenches were superior to sprays, and optimal drench concentrations were 1,000–2,000 mg·L^{-1} BA.

CONCLUSION

The results from this study demonstrate that BA drenches enhanced axillary shoot production, and consequently improved pot fill for 'June' hosta plants, without negative side-effects of phytotoxicity. Further testing of additional hosta cultivars should be conducted with BA drenches to determine optimal concentrations.

LITERATURE CITED

Keever, G.J. 1994. BA-induced offset formation in hosta. J. Environ. Horticulture 12(1):36-39.

Garner, J.M., G.J. Keever, D.J. Eakes, and J.R. Kessler. 1997. BA application promotes offset formation in hosta cultivars. J. Environ. Horticulture 15(3):119-122.

Schultz, H.C., G.J. Keever, J.R. Kessler Jr, and R.R. Dute. 2000. Benzyladenine improves summer quality of hosta. J. Environ. Horticulture 18(1):49-52.

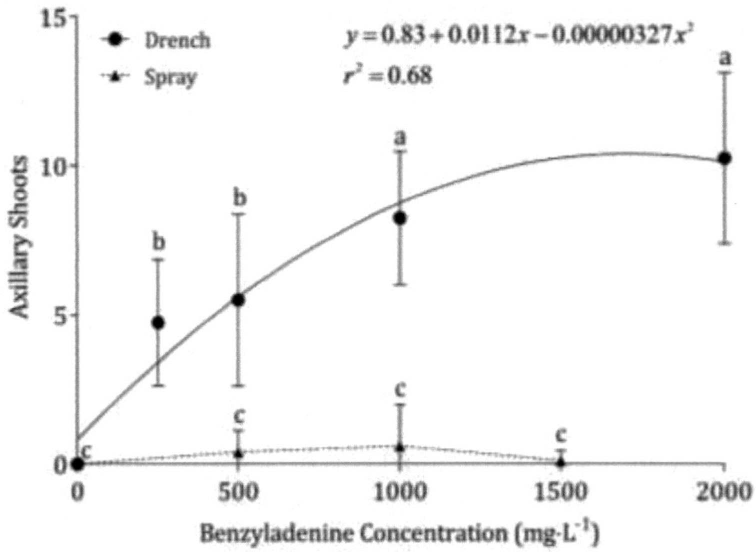

Figure 1. Hosta shoot development eight weeks after BA application. Significant differences among all treatments are indicated by lower case letters with $P < 0.05$.

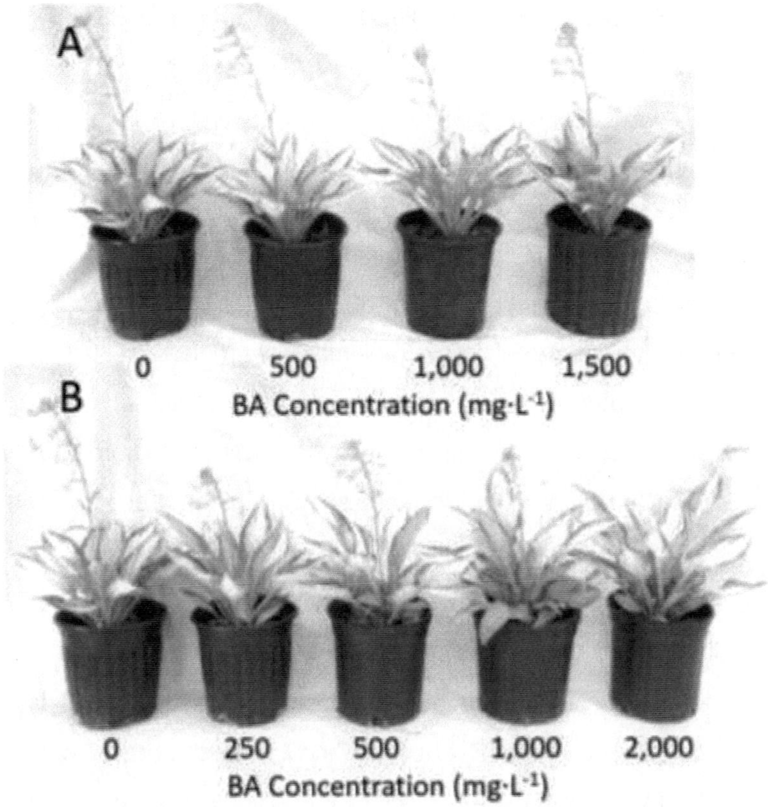

Figure 2. Hosta plants that received (A) two sprays or (B) one drench of BA, eight weeks after application.

TANK MIXING CONFIGURE (BA) WITH PLANT GROWTH RETARDANTS IMPROVES BRANCHING

Joyce Latimer, Margaret Aiken, John Freeborn and Holly Scoggins
Department of Horticulture, Virginia Tech, Blacksburg, VA 24061

Tank mixes are generally designed to provide a synergistic effect. In other words, you can use lower rates of both products to get the same or greater efficacy. Tank mixes of Configure (600 ppm) with Dazide (2500 ppm; daminozide, Fine Americas, Inc.) or the proprietary carrier used in Dazide were applied to liners just after removal from mist when plants were moderately rooted. Only in Agastache 'Tutti Frutti' did the tank mix, Configure + Dazide, increase the number of branches over that of Configure alone. However, in both Rosmarinus 'Hill Hardy' and Gaura 'Whirling Butterflies,' while Dazide alone reduced branching, the Tank Mix restored the number of branches to the same as untreated control plants. Tank mix combinations of Configure (600 ppm) and Piccolo (120 ppm; paclobutrazol, Fine Americas) were applied to commercially grown liners at about 10 days after potting. In Echinacea 'Doubledecker' and Heuchera 'Silver Lode,' Piccolo reduced branching, which was restored in plants receiving the Tank Mix. In Lobelia speciosa 'Fan Deep Rose,' Configure alone or the Tank Mix increased early branching relative to control plants or those treated with Piccolo alone. Tank mixes of Configure (600 ppm) and Concise (60 ppm; Fine Americas) applied to commercially produced liners tended to reduce branching of Verbena 'Homestead Red Carpet' while Concise alone caused no reduction in branching. In summary, we found no consistent synergism with these tank mixes, but generally found a positive influence on plant branching when Configure was tank mixed with these plant growth retardants.

SMOOTH BROME (BROMUS INERMIS) GROWTH RESPONSE TO RYZUP® SMARTGRASS APPLICATION IS AFFECTED BY SURFACTANT

Michael D. Rethwisch, Craig J. Hruska, Garrett Tooker, and Kaleb Scheffler
University of Nebraska-Lincoln, Butler County Extension, 451 N. 5th Street, David City, NE, USA 68632-1666 mrethwisch2@unl.edu (402) 367-7410

In 2012, striking differences were noted in adjacent field test plots of smooth brome (*Bromus inermis*) treated with RyzUp® SmartGrass (active ingredient = gibberellic acid 3). Subsequent investigation noted different surfactants were used between the two sites, providing the first clue that the correct surfactant was critical to crop response to gibberellic acid application. A number of surfactants were screened in a series of RyzUp® SmartGrass applications to smooth brome in small plot field trials over the next several years to determine which surfactants were most effective. Natural forage height and extended leaf measurements indicated significant differences due to surfactant included, and that best and most consistent growth responses occurred when non-ionic surfactants which contained ammonium sulfate were used with gibberellic acid. These combinations resulted in positive economic hay production values across several nitrogen fertility levels in fields, at 30 days and especially at 60 days post treatment. It is thought that the ammonium component of these surfactants help to keep the gibberellic acid from being denatured in the spray tank, while the additional nitrogen is also beneficial.

INTRODUCTION

Initial experiments in 2012 with RyzUp® SmartGrass 40 WDG (Valent USA, active ingredient = gibberellic acid 3 (GA-3)) on smooth brome, *Bromus inermis* L., had resulted in widely contrasting results. One trial resulted in easily discernable growth differences in small plot trials (Fig. 1), but a second study in the same field involving three RyzUp® SmartGrass 40 WDG rates had in very little growth being noted except in tractor wheel tracks (Fig. 2).

The opposing results were the first clue that surfactants and their associated additives are critical to successfully increasing forage production of smooth brome with GA-3 in the form of RyzUp® SmartGrass 40 WDG. This also indicated that there may be differences between the non-ionic surfactant (NIS) product class (product label notes to use a NIS surfactant).

As data for interactions of RyzUp® SmartGrass 40 WDG and various surfactants for smooth brome production do not exist, this experiment was initiated to document smooth brome growth response to RyzUp® SmartGrass 40 WDG to various surfactants.

MATERIALS

Six surfactant products were used with RyzUp SmartGrass which was applied at 0.3 oz./acre at 28 gpa solution to smooth brome at three field sites with three fertility levels (low, medium, high) in Butler County during spring 2015. This negated the potential growth effect due to nitrogen in some products.

SURFACTANT INFORMATION

NON-IONIC SURFACTANTS: Organic BioLink® Spreader-Sticker (Westbridge Agricultural Products, Vista, CA) consists of 10.1% soapbark, alkylphenol ethoxylate, and polysaccharide. Use rate was 8 fl. oz. per 100 gallons (3.36 fl. oz./acre).

Two (2) rates of AgriSolutionsTM Class Act® NG® (distributed by Winfield Solutions, LLC, St. Paul, MN) were evaluated with RyzUp® SmartGrass (2.5% or 1.25% v/v). Class Act® NG® is a water conditioning agent/non-ionic surfactant (NIS) blend product consisting of 50.5% ammonium sulfate, corn syrup and alkyl polyglucoside, and 49.5% constituents ineffective as spray adjuvants. This formulation contains 1.3 lbs. of sugar and 3.4 lbs. of dry ammonium sulfate/gallon. Actual rates/acre in this experiment at 2.5% v/v were 0.7 gallons/acre (equivalent to 2.38 lbs./acre of ammonium sulfate, or 0.5 lbs. of actual nitrogen and 0.57 lbs./acre of sulfur), and one-half of these amounts for the 1.25% v/v rate.

Fastrack® Water Conditioning and Surfactant Adjuvant System (United Suppliers, Inc., Eldora, Iowa) is a 100% proprietary blend of water conditioning agents, ammonical nitrogen and non-ionic surfactant. This ammonium sulfate replacement product is a mixture of potassium salts of hydroxyl carboxylic, and phosphoric acid, alkyl amine alkoxylate, alkyl polyoxyethylene phosphate, and urea ammonium nitrate (62%) with 38% constituents ineffective as spray adjuvants. It has 26% surfactant and was used at 0.5% v/v.

UltraSurf™ AMS (United Suppliers, Inc., Eldora, Iowa) is a pre-mixed formulation of ammonium sulfate solution (80%) and corn based non-ionic surfactant (20%), and is a water conditioner. surfactant It consists of ammonium sulfate, polyols, alkyl polyoxyethylene sulfate and aryl sulfonate (53.9%), and 46.1% constituents Ineffective as spray adjuvants. Surfactant concentration is 10%. Use rate was 2.5% v/v.

PENETRANT AND WETTING AGENT: WetCit (Oro Agri Inc., Fresno, CA) is designed to enhance the efficacy of plant growth regulators and other agrichemicals. It consists of 8.15% alcohol ethoxylate, with 91.85% constituents ineffective as spray adjuvants. Use rate was 0.25% v/v.

OTHER: Aegis ESR (marketed by AMGI LLC, Taft, TX; product previously known as AX-321, Atonik, and Chaperone) is a protein enhancer that increases the utilization of proteins necessary for plant growth. Active ingredients in this product consist of 0.3% sodium p-nitrophenolate, 0.2% sodium o-nitrophenolate, and 0.1% sodium 5-nitroguaiacolate. While typically not considered to be a surfactant, it previously had excellent local results when used in combination with RyzUp® SmartGrass and a non-ionic surfactant for local oat forage production, but had not been evaluated with RyzUp® SmartGrass by itself. Use rate was 10 oz./acre.

METHODS

Treatments were applied using a ShurFlo 600 back-pack sprayer equipped with a boom containing four Lurmark 04F80 flat fan nozzles. Treatment design was a randomized complete block utilizing four replications for each treatment at each of the three sites. Plots were 7 ft. x 25 ft.

Plant measurements of standing heights and extended leaf heights were obtained on roughly a weekly basis throughout the experiment. Five (or more) plants were sampled in each plot and data recorded. Yield data were obtained at approximately 60 days after application by cutting three 1.125 squares from each plot, oven drying, weighing and calculating hay yields per plot.

Data were analyzed and treatment means statistically separated using the Tukey-Kramer Honestly Significant Difference (HSD) test (JMP 10.0.0, SAS Institute Incorporated, Cary, NC).

RESULTS

Easily discernable visible differences in smooth brome growth were noted at all locations within three weeks of application (Fig. 3).

Windy conditions on sample days often affected natural forage heights measurements, thus extended leaf heights were a more reliable indicator of actual results. While multiple statistical differences were noted between treatments at each site (not shown) for early season extended leaf measurements, a highly significant growth increase in the combined data analysis for mean site differences ($p<0.003$) was noted only for NIS water conditioning surfactants that contained ammonium sulfate and used at a minimum rate of 1.25% v/v.

Fertility at the three sites greatly affected the smooth brome hay yield. There were no statistical differences at any site or across locations for 30 (Fig. 4) or 60 days post treatment due to yield variations. Positive net economics were noted at the medium and high fertility sites, following the trend noted for early season growth associated with NIS surfactants.

CONCLUSIONS

Consistent growth responses occurred when water conditioning non-ionic surfactants containing ammonium sulfate were used with GA-3. These combinations resulted in positive economic hay production values across several nitrogen fertility levels in fields, at 30 days and especially at 60 days post treatment. The ammonium component of these surfactants may help gibberellic acid from being denatured in the spray tank, while the additional nitrogen is also beneficial.

Figs. 1 (left) and 2 (right). Small plots with highly visible growth differences associated with application of RyzUp® SmartGrass 40WDG to smooth brome. Growth response noted primarily in tractor tire tracks.

Fig. 3. Smooth brome growth at 13 days post treatment with 0.3 oz./acre of RyzUp® SmartGrass 40 WDG and various surfactants at the low fertility site.

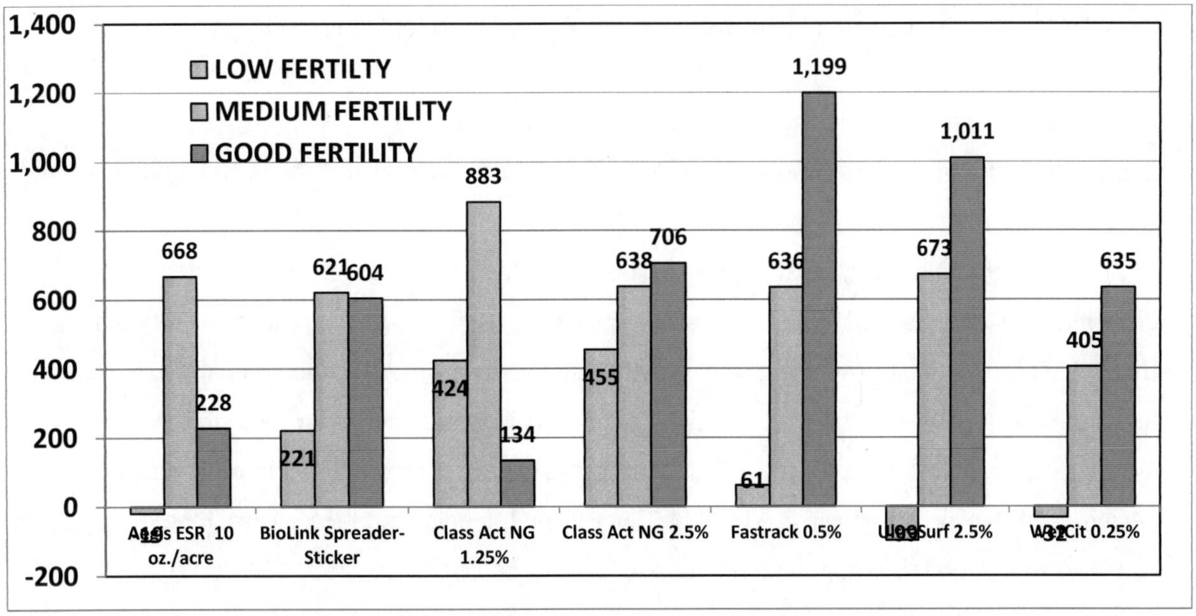

Fig. 4. Mean smooth brome hay yields (lbs./acre) differences from RyzUp® SmartGrass 40 WDG used with various surfactants across three fertility levels at 30 days post treatment

COMPARISON OF FOLIAR SPRAYS FOR COLEUS GROWTH CONTROL

Paul Cockson, Josh B. Henry, Ingram McCall and Brian E. Whipker
Department of Horticultural Science, North Carolina State University, Raleigh, NC 27695 U.S.A.

A number of plant growth regulators (PGR) are recommended for use in controlling the growth dynamics of coleus [Plectranthus scutellarioides (L.)]. This study compared the efficacy of three PGR products with the end goal of providing growers with more refined recommendations to grow fuller and more robust plants. Cuttings of 'Stained Glassworks Molten Lava' and 'Wall Street' were transplanted into 5-inch pots. PGR foliar sprays were applied on 26 April: paclobutrazol (Piccolo 10 XC, Fine Americas) at 10, 20, 40, or 60 mg·L^{-1}; uniconazole (Concise, Fine Americas) at 5, 10, 20 or 40 mg·L^{-1}; and ethephon (Collate, Fine Americas) at 125, 250, 500, or 1000 mg·L^{-1}. At the concentrations used, paclobutrazol provided a limited amount of control of plant growth. Uniconazole at 20 and 40 mg·L^{-1} limited internodal plant stretch whereby producing a more compact plant. Ethephon provided the highest degree of plant growth control and resulted in the most compact plants. While cultivar response did vary, ethephon foliar sprays were the most effective method of growth control for coleus. The use of ethephon offers growers a valuable tool to produce fuller and more compact consumer products.

INTRODUCTION

Coleus (Plectranthus scutellarioides (L.)) are grown for their brightly colored foliage (Nau, 2011). Many of the vegetatively propagated cultivars are vigorous growers and require plant growth regulators (PGRs) to control their dimensions. Whipker (2017) lists a number of PGRs that are suitable for use on coleus. One option is daminozide at 2,500 to 4,000 mg·L^{-1} plus chlormequat chloride at 1,000 to 1,500 mg·L^{-1}. Chlormequat chloride alone at 800 to 1,500 mg·L^{-1} can also be used. In addition, paclobutrazol at 5 to 30 mg·L^{-1}, uniconazole at 5 to 20 mg·L^{-1}, or ethephon at 500 mg·L^{-1} can be used to control growth. This study was conducted to compare the efficacy of ethephon, paclobutrazol and uniconazole sprays on coleus in order to provide useful guidelines for growers.

MATERIALS AND METHODS

'Stained Glassworks Molten Lava' and 'Wall Street' coleus cuttings (Dümmen Orange, Columbus, Ohio) were stuck on 14 March 2017 into 128 cell flats (3.8 tall x 2.4 x 2.4 cm) filled with

Sunshine Mix #1 (Sun Gro Horticulture, Agawam, MA). The plants were rooted in a glass greenhouse at 35°N latitude in Raleigh, NC and allowed to establish. The rooted cuttings were transplanted on 4 April into 5-inch (13 cm) diameter pots (Dillen, Middlefield, OH) filled with Sunshine Mix #1 (Sun Gro Horticulture, Agawam, MA) and moved into a polyethylene covered greenhouse for the remainder of the experiment. The plants were fertilized at every irrigation with 13N−0.86P−10.8K (Excel® 13−2−13, The Scotts Co., Marysville, Ohio) mixed at 150 mg·L-1 nitrogen.

The experiment was completely randomized with six single plant replicates of each of the 13 treatments. Coleus plants were treated on 26 April with a spray of 125, 250, 500, or 1,000 mg·L^{-1} ethephon (Collate; Fine Americas), 10, 20, 40 or 60 ppm paclobutrazol (Piccolo 10XC; Fine Americas) or; uniconazole (Concise; Fine Americas) at 5, 10, 20 or 40 mg·L^{-1}. The foliar sprays were applied at the rate of 203.7 mL·m^{-2}, and were sprayed to the point of slight foliar runoff.

The experiment was terminated on 22 May. Data for coleus plant height and diameter were analyzed using SAS (version 9.4; SAS Institute, Cary, NC). Data were subjected to PROC GLM and the least squared means were separated by Tukey's honestly significant differences at P≤0.05.

RESULTS

There were no symptoms of phytotoxicity with any of the three PGR chemicals or concentrations used. Overall, there was a cultivar times chemical interaction. The cultivars had different responses to the PGRs. The older cultivar 'Stained Glassworks Molten Lava' (Fig. 1) was less responsive to the PGRs and concentrations used than 'Wall Street' (Fig. 2). 'Wall Street' is a recent introduction and was bred for more compact plants.

Plant height and diameter of 'Stained Glassworks Molten Lava' was similar across all the paclobutrazol concentrations trialed (0 to 60 mg·L^{-1}) (Fig 1A). Either higher concentrations or substrate drenches should be trialed to determine suitable rates.

Uniconazole foliar sprays also did not control plant height or diameter of 'Stained Glassworks Molten Lava' plants (Fig 1B). In the case of 10 mg·L^{-1} and 20 mg·L^{-1} uniconazole, plant height and plant diameter, respectively, were actually greater than the untreated controls.

Growth control of 'Stained Glassworks Molten Lava' occurred with the highest concentration of ethephon (Fig 1C). Plants treated with 1000 mg·L^{-1} ethephon were 18.7% shorter and 15% more slender when compared to the untreated controls.

'Wall Street' was responsive to all the PGRs trialed (Fig. 2). With paclobutrazol foliar sprays, plant height was significantly controlled by 10.4% at >20 mg·L^{-1} as compared with the untreated control (Fig. 2A). With the control of plant height with 20 mg·L^{-1} paclobutrazol, this resulted in a corresponding 11% increase in plant diameter as compared with the untreated plants.

Uniconazole was more effective in controlling both plant height and diameter of 'Wall Street' coleus (Fig. 2B). Plant height was significantly shorter (14.2%) with 10 mg·L^{-1} uniconazole, as compared with the untreated control. Plant diameter was wider than the untreated control with 5 mg·L^{-1}, but was smaller with 40 mg·L^{-1}.

'Wall Street' was most responsive to ethephon. Plant height was 20% shorter with 125 mg·L^{-1} ethephon, as compared with the untreated control. Plant diameter was not affected until the ethephon concentration was at 1,000 mg·L^{-1}.

CONCLUSION

'Stained Glassworks Molten Lava' was less responsive to the PGR foliar sprays and concentrations used when compared with 'Wall Street'. Only a foliar spray of 1000 mg·L^{-1} ethephon was effective producing plants that were 18.7% shorter and 15% slimmer than the untreated controls. 'Stained Glassworks Molten Lava' has been commercially available for a number of years. It is considered to be a vigorous grower and consequently, higher concentrations of PGRs should be trialed as foliar sprays. Alternatively, substrate drenches may be a more effective method to use to control growth.

'Wall Street' was responsive to all the PGRs trialed, with ethephon having the most affect on plant growth. Because ethephon foliar sprays are also a cost effective method to use for growth control, it should be the preferred method to use to control plant stretch of coleus.

LITERATURE CITED

Nau, J. Ball Redbook, vol.2, ed. 18. 2011. Ball Publishing, West Chicago, IL. pp. 785.

Whipker, B.E. Plant Growth Regulators for Annuals. 2017. GrowerTalks, Ball Publishing, West Chicago, IL. pp. 52.

A 'Stained Glassworks Molten Lava'

Paclobutrazol Foliar Spray

Parameter	Concentration (mg/L)					P value[z]
	0	10	20	40	60	
Height (cm)	26.8	27.6	28.7	27.1	28.0	NS
Diameter (cm)	36.6	38.9	36.2	39.9	38.0	NS

B

Uniconazole Foliar Spray

Parameter	Concentration (mg/L)					P value[z]
	0	5	10	20	40	
Height (cm)	26.8 bc	29.3 ab	32.3 a	27.2 bc	25.5 c	**
Diameter (cm)	36.6 abc	40.2 a	38.7 ab	34.4 c	35.4 bc	*

C

Ethephon Foliar Spray

Parameter	Concentration (mg/L)					P value[z]
	0	125	250	500	1000	
Height (cm)	26.8 ab	28.6 a	29.3 a	25.1 b	21.8 c	***
Diameter (cm)	36.6 a	35.3 a	37.1 a	35.6 a	31.0 b	*

Figure 1. The effects of (A) paclobutrazol, (B) uniconazole, and (C) ethephon foliar sprays on growth control of 'Stained Glassworks Molten Lava' coleus (Plectranthus scutellarioides).

A **'Wall Street'**

Paclobutrazol Foliar Spray

0 10 20 40 60 mg/L

Parameter	Concentration (mg/L)					P value[1]
	0	10	20	40	60	
Height (cm)	26.0 a	24.1 ab	23.3 bc	22.8 bc	20.7 c	**
Diameter (cm)	36.1 bc	36.8 abc	40.1 a	39.6 ab	34.0 c	*

B

Uniconazole Foliar Spray

0 5 10 20 40 mg/L

Parameter	Concentration (mg/L)					P value[1]
	0	5	10	20	40	
Height (cm)	26.0 a	25.8 a	22.3 b	19.9 b	15.4 c	***
Diameter (cm)	36.1 b	43.1 a	33.7 b	33.1 b	26.3 c	***

C

Ethephon Foliar Spray

0 125 250 500 1000 mg/L

Parameter	Concentration (mg/L)					P value[1]
	0	125	250	500	1000	
Height (cm)	26.0 a	20.8 b	20.8 b	19.3 c	18.6 c	***
Diameter (cm)	36.1 a	34.6 a	36.0 a	33.5 a	30.4 b	**

Figure 2. The effects of (A) paclobutrazol, (B) uniconazole, and (C) ethephon foliar sprays on growth control of 'Wall Street' coleus (Plectranthus scutellarioides).

INTERACTIVE EFFECTS OF ABSCISIC ACID (ABA) AND TEMPERATURES ON SKIN ANTHOCYANINS IN CABERNET SAUVIGNON BERRIES (*VITIS VINIFERA* L.)

Sanliang Gu

Viticulture and Enology Research Center, California State University, Fresno, 2360 E. Barstow Avenue, Fresno, CA 93740, USA

Anthocyanins are one of the major quality components in red winegrapes. Temperatures greater than 30 °C inhibit synthesis and increase degradation. Abscisic acid (ABA) has been used as a vineyard management tool to enhance fruit anthocyanins and skin color for table grape production in warm regions. Variable effectiveness of ABA at different stages of fruit ripening suggests that berries may vary in their sensitivity to ABA application. Experiments were conducted to examine the interactive effects of ABA and temperatures on the accumulation of anthocyanins in Cabernet Sauvignon winegrape berries at various stages of fruit ripening. Soft green, pink, red berries at veraison and purple berries two weeks prior to harvest were detached from the rachis with pedicels intact, submerged in 0 or 600 $mg \cdot L^{-1}$ ABA solution for 10 minutes, placed in plastic bags with small holes, incubated at 10, 20, 30, and 40 °C for 4 days, and then analyzed for delphinidin (De), cyanidin (Cy), petunidin (Pt), peonidin (Pn), and malvidin (Mv) -3-glucosides (3-G), -3-acetyle-glucosides (3-A-G), and -3-coumaroyl-glucosides (3-C-G). Depending on berry ripening stage, temperatures interacted with ABA to affect the composition and total content of anthocyanins. Without ABA, the optimum temperatures for accumulation of anthocyanins were 20 and 30 °C. Compared to the untreated control, ABA-treated pink and red berries accumulated more anthocyanins at 20, 30, and 40 °C while greater anthocyanin content in ABA-treated purple berries was observed only at 30 °C. As skin anthocyanins increased during fruit ripening, all forms of De-, Pt-, Pn- and Mv-3-G increased, but all forms of Cy-3-G as well as De-, Pt-, Pn- and Cy-3-C-G were at very low concentration or did not exist. The soft green and pink berries did not accumulate non Mv based anthocyanins at 10 °C after 4 days, either with or without ABA. Highest levels were observed for De-, Pt-, Pn-3-G and Pn-3-A-G at 20 °C, Mv-3-G and Mv-3-A-G at 30 °C, and Mv-3-C-G at 40 °C, with or without ABA. ABA enhanced the accumulation of Mv based anthocyanins at 30 and 40 °C as well as 3-G based anthocyanins at 20, 30 and 40 °C. The research suggests that all skin anthocyanins were more sensitive to ABA at high temperatures at berry coloring while non Mv based anthocyanins had a greater degradation potential.

LIVING GREENWALLS: SUBSTRATE FEEDBACK IN A MODEL PLANT

Mikayla Mack[1] and Sonja L. Maki[2]
Department of Biology[1] and Department of Plant and Earth Sciences[2]
410 South Third Street, University of Wisconsin-River Falls, River Falls, WI 54022

Living greenwalls are interior and exterior features that can be found incorporated into biophilic designs in businesses, industry, educational and home landscapes. In addition, living greenwalls are also used for processes such as air and water purification systems. In a living greenwall, the roots of plants are held in a vertical orientation by a number of different methods. Plants are watered and fertilized hydroponically. The objective of this study was to examine the growth of the model plant Chamaecrista fasciculata in a large, established classroom greenwall and in test modules located in a greenhouse setting. Several substrates and transplanting methods (seeding, bare root, and plugs) were assessed and environmental parameters (pH, electrical conductivity (EC), dissolved oxygen (DO)) and irradiance levels were also monitored. In addition, we also studied the expression of a nonsymbiotic hemoglobin gene in silico in as a potential marker for root health in foam based greenwall systems which are subjected to wet and drying cycles.

GIBBERELLIC ACID EFFECTS ON AGRONOMIC CROPS IN TENNESSEE

T. C. Mueller[1] and J. R. Cranmer[2]

[1]Department of Plant Sciences, The University of Tennessee, Knoxville, TN, 37996-4561 USA

[2]FMD Specialist, Valent USA LLC, Morrisville, NC 27560 USA

Gibberellin is a naturally occurring plant growth regulator known to cause multiple responses in plants. Possible effects include growth stimulation, improved dry matter yields, and better stress tolerance of environmental challenges. Ryzup Smartgrass (RS) is a 40% active w:w gibberellin product that can be applied in most early season pesticide and fertilizer programs and is approved for use in conventional and organic silage corn production.

Field plots were established and conducted in 2014 and 2015 in Knoxville, TN (latitude 35.971438, longitude -83.854094) on the East TN Research and Education Center. Experiments included winter wheat (1), corn (3) and soybeans (1) grown using typical production practices for the area. RS was applied at early vegetative stages to each of the crops and a variety of plant measurements was taken in each study. Application of RS was made using small plot equipment in a carrier volume of 190 L/ha of water carrier at a pressure of 275 kPa using flat fan nozzles to provide complete surface coverage. For ease of discussion, the data were pooled within each study by those treatments not receiving RS and comparing all treatments within that study that received an RS application.

Wheat vigor scores 12 days after treatment were increased and yield was increased 164 kg/ha. Corn yield was increased in 2 of 3 studies, with changes being -222, + 370 and + 148 kg/ha. Soybeans showed no change in yield in the single study in this report. In all studies, there was no crop phyto-toxicity observed from any application, even when applied with a broad range of herbicides, fungicides and insecticides. Application with micro-nutrients also did not cause any apparent crop injury.

TESTING OF NEW CLONING GEL FOR PLANT CUTTINGS ROOTING

Valentin Savov[1], Konstantin Chakalov[2], Todorka Popova[2], and Gergana Angelova[2]
[1]Sofia University "St. Kliment Ohridski", Bulgaria
[2]Balkan Plant Sciences Ltd., Bulgaria

Due to the poor solubility of phytohormones in water, auxin-containing powders are a typical choice when trying to root plant cuttings. However, plants can absorb a very small part of the auxins in the powders per unit of time, and, furthermore, much of them is washed away from the cuttings during irrigation. That's the reason for the poor effect of powders as stimulators of root formation.

The aim of the study is to examine the effect of a rooting product available in dry form which dissolves easily to a gel state, as a means of activating the root formation and stimulating the growth of the newly formed roots of softwood and semi-hardwood cuttings regardless of the plant species. For this purpose, natural or synthetic auxins are combined with phlorizin and indigo carmine. Amorphophallus konjac extract is used as a gelling agent. Phlorizin has an inhibitory effect on the indolyl acetate oxidase activity in plant roots.

The experiments include three types of auxins - naphthylacetic acid (NAA), indoleacetic acid (IAA), and indolebutyric acid (IBA). The dose-response curve has two peak values for IAA and NAA - 250 and 750 mg·L^{-1}, and 4 for IBA - 550, 950, 1250, and 1750 mg·L^{-1}, on the background of the other components. In the case of IBA, however, the higher two doses provide 100% rooting, and the lower ones - 85-90%. Depending on the species, the first roots occur from the 5th to the 15th day of the beginning of the process.

COMPARISON OF ORNAMENTAL PEPPER RIPENING ENHANCEMENT WITH ETHEPHON APPLIED AS A SPRAY OR DRENCH

Josh B. Henry, Brian E. Whipker, and Ingram McCall
Department of Horticultural Science, North Carolina State University, Raleigh, NC 27606 U.S.A.

Ethephon (Collate; Fine Americas) can be used to enhance the ripening of fruits including ornamental peppers. Previous work found ethephon sprays to be an effective method of ripening enhancement; however, ethephon sprays also led to significant fruit abscission and phytotoxicity to the foliage. This study compared the efficacy of ethephon drenches and sprays on an ornamental pepper cultivar (Capsicum annuum 'Tango Red'). Greenhouse grown pepper plants were treated with ethephon when ~75% of the fruit had reached mature size but had not begun to ripen. These plants received either a spray or a drench of 75, 150, 300, or 600 mg·L^{-1} ethephon. The study was terminated three weeks after the ethephon application, and the number of green, blushing, red, and abscised peppers was totaled. A phytotoxicity rating was also conducted based on foliar symptoms. All plants treated with an ethephon drench and all but one ethephon spray concentration had more red peppers than the untreated control. Plants that received an ethephon drench of 75 or 150 mg·L^{-1} had higher fruit set compared to untreated plants or plants that received an ethephon spray. All plants receiving a spray had higher levels of phytotoxicity than plants receiving a drench, while drenched plants had low phytotoxicity ratings or were asymptomatic similar to the untreated controls. In this study, the optimal concentrations of 150 – 300 mg·L^{-1} ethephon applied as a drench resulted in higher fruit set and lower incidence of phytotoxicity.

INTRODUCTION

Ornamental peppers (Capsicum annuum) are grown for the abundance of brightly colored fruit they produce upon maturation. The immature fruit are typically green, but over time can ripen to vibrant yellows, reds, or a variety of other colors. Fruit ripening can be enhanced with applications of ethephon (2-Chloroethylphosphonic acid). Previous studies investigated the effect of ethephon sprays on fruit abscission (Beaudry and Kays, 1988) and ripening of edible peppers (Batal and Granberry, 1982; Cooksey et. al, 1994; Mao and Motsenbocker, 2002). In these studies, ethephon increased fruit ripening but also caused greater fruit abscission when applied at higher concentrations (Cooksey et. al, 1994). Fruit abscission is sometimes beneficial for harvest of edible peppers (Mao and Motsenbocker, 2002), but is not desirable for ornamental pepper production. Additionally, phytotoxicity and leaf abscission may occur on pepper plants sprayed with ethephon (Armitage, 1989), rendering the plants unmarketable.

Although ethephon is typically applied as a foliar spray, ethephon substrate drenches have been of recent interest for floriculture crop production (Miller et. al, 2012). There is potential to avoid the phytotoxicity symptoms associated with ethephon sprays by applying ethephon to the substrate instead of the foliage. Miller et. al (2012) demonstrated that ethephon drenches successfully suppressed plant height without phytotoxicity symptoms developing on the foliage. Applying ethephon as a substrate drench may provide a safer alternative for ornamental pepper growers to enhance ripening and reduce production time. This study was conducted to compare the efficacy of ethephon drenches and sprays on ornamental peppers.

MATERIALS AND METHODS

'Tango Red' ornamental pepper seeds (Fred C. Gloeckner & Co., Inc., Harrison, NY), were sown on 6 April 2016 into 1204 flat inserts with cell dimensions of 5.7 x 3.8 x 5.4 cm and propagated in a glass greenhouse at 35°N latitude in Raleigh, NC. The seedlings were transplanted into 5.5-inch diameter pots (Dillen, Middlefield, OH) with dimensions of 14.0 x 11.4 x 11.1 cm and moved into a polyethylene covered greenhouse for the remainder of the experiment. The plants were fertilized at every irrigation with 13N–0.9P–10.8K mixed at 150 $mg{\cdot}L^{-1}$ nitrogen.

The experiment was completely randomized with nine single plant replicates of each of the nine treatments. Pepper plants were treated on 3 August 2016 with either a spray or a drench of 75, 150, 300, or 600 $mg{\cdot}L^{-1}$ ethephon (Collate; Fine Americas) when ≥75% of the peppers had reached mature size but were still green. Sprays were applied to the foliage and fruit at a rate of 203.7 $mL{\cdot}m^{-2}$, and were sprayed to the point of slight runoff. Drenched plants had 88.7 mL of solution applied to the substrate of each pot.

The experiment was terminated 21 days after treatment on 24 August. Peppers were removed from the plants, and the number of green, blushing, red, and abscised peppers was totaled. A phytotoxicity rating was also conducted based on foliar phytotoxicity symptoms. The scale was 0 – 3 with 0 being no damage and 3 being high levels of damage. The highest level of damage consisted of severe chlorosis of the foliage, leaf abscission, and some purple spotting on the leaves. The peppers and remaining shoot were then dried for at least 72 hours at 70 °C and used to obtain the dry mass.

Data for pepper coloration, abscission, dry mass, and foliar phytotoxicity were analyzed using SAS (version 9.4; SAS Institute, Cary, NC). Data were subjected to PROC GLM and the least squared means were separated by Tukey's honestly significant differences at $P \leq 0.05$.

RESULTS

Whether a plant was treated with ethephon and the method by which ethephon was applied led to several differences compared to untreated plants. Sprays of 150 – 600 mg·L^{-1} ethephon led to significant levels of fruit abscission, but drenches did not (Table 1). Plants sprayed with 75, 150, and 600 mg·L^{-1} ethephon had more red peppers than untreated plants, while all drench concentrations resulted in higher numbers of red peppers (Table 1). All spray concentrations resulted in fewer green peppers than control plants; however, drenches of 75 – 300 mg·L^{-1} ethephon had similar numbers of green peppers as control plants (Table 1). Although sprays of 150 – 600 mg·L^{-1} ethephon led to significant levels of fruit abscission, all sprayed plants had similar numbers of total fruit as untreated plants (Table 1). Conversely, plants drenched with 150 – 300 mg·L^{-1} ethephon had significantly more peppers than control plants (Table 1).

The total quantity of pepper dry mass was also significantly affected by ethephon treatment. All plants that received an ethephon drench had similar total pepper dry mass to untreated plants (Fig. 1). All plants treated with a spray had lower pepper dry mass compared to untreated plants (Fig. 1). Plants that received a drench had phytotoxicity ratings that were not significantly greater than that of the asymptomatic control (Fig. 2). In fact, the majority (>50%) of drenched plants were completely asymptomatic. The only symptoms that occurred on drenched plants included some chlorotic spotting on the foliage. The phytotoxicity rating increased with increasing spray concentrations, with uniformly high symptoms observed on all plants (Fig. 2). Phytotoxicity symptoms were present on all sprayed plants (Fig. 3).

DISCUSSION

One of the most unexpected findings from this study was that moderate concentrations of ethephon applied as a drench appeared to enhance not only the number of red fruit, but also the total number of fruit compared to untreated plants. This indicates that ethephon drenches may stimulate fruit set. A benefit to consumers is that ornamental pepper plants drenched with ethephon may have more red fruit at the time of purchase, with additional green and blushing fruits to potentially extend the color for a greater period of time. The optimal ethephon drench concentrations determined from this study were 150 – 300 mg·L^{-1}. These concentrations did not result in significant fruit abscission or phytotoxicity to the foliage and led to higher numbers of red peppers with increased overall fruit set. This indicates that ethephon drenches can be successfully used by growers to produce ornamental peppers and may be a superior alternative to ethephon sprays.

Previous work investigating ethephon substrate drenches found some negative side effects of drenches. For instance, Miller et. al (2012) found that ethephon drenches led to a delay of flowering in several bedding plant species. Ethephon drenches were also found to cause flower bud abortion and premature floral senescence in some species. None of these issues would be problematic for ornamental pepper production, as ethephon applications are made after anthesis. Additionally, because Miller et. al (2012) did not observe any foliar phytotoxicity symptoms, and the symptoms observed on the drenched plants in our study were minimal, it may be concluded that ethephon drenches maintain many of the desired effects of an ethephon spray without the same magnitude of undesirable side effects.

Using ethephon enables ornamental pepper growers to reduce the total crop time by enhancing fruit ripening (Armitage, 1989). Ethephon drenches resulted in the development of similar numbers of red peppers as ethephon sprays, without the severity of phytotoxicity and fruit abscission. Further studies should investigate drench application timing to optimize ethephon treatments and develop recommendations that may be applied to commercial production practices. There are currently no ethephon products registered for use in greenhouses as a drench. Therefore, these results should be considered experimental.

LITERATURE CITED

Armitage, A.M. 1989. Promotion of fruit ripening of ornamental peppers by ethephon. HortScience 24(6):962-964.

Batal, K.M. and D.M. Granberry. 1982. Effects of growth regulators on ripening and abscission of pimiento and paprika peppers. HortScience 17(6):944-946.

Beaudry, R.M. and S.J. Kays. 1988. Effect of ethylene source on abscission of pepper plant organs. HortScience 23(4):742-744.

Cooksey, J.R., J.E. Motes, and B.A. Kahn. 1994. Calcium and ethephon effects on paprika pepper fruit retention and fruit color development. HortScience 29(7):792-794.

Mao, C. and C.E. Motsenbocker. 2002. Effects of ethephon on tabasco pepper fruit ripening and abscission at the fruit-receptacle junction. Scientia Horticulturae 93(3):357-365.

Miller, W.B., N.S. Mattson, X. Xie, D. Xu, C.J. Currey, K.L. Clemens, R.G. Lopez, M. Olrich, and E.S. Runkle. 2012. Ethephon substrate drenches inhibit stem extension of floriculture crops. HortScience 47(9):1312-1319.

Table 1. Mean number of peppers of each color from each ethephon treatment, as well as the number of abscised peppers.

Application method	Ethephon concentration ($mg \cdot L^{-1}$)	Number of peppers				
		Total	Green	Blushing	Red	Abscised
Control	0	38.1bc[1]	15.1a	7.7abc	15.3b	0d
Spray	75	34.7c	2.1b	5.6bcd	24.2a	2.8bcd
	150	36.4c	0.9b	4.4cd	25.2a	5.9a
	300	30.4c	0.3b	4.4cd	20.8ab	4.9ab
	600	31.0c	0.1b	2.4d	25.2a	3.2abc
Drench	75	45.8ab	13.7a	9.6ab	22.6a	0d
	150	47.4a	11.4a	9.9ab	25.0a	1.1cd
	300	49.4a	10.9a	11.1a	25.9a	1.6cd
	600	37.4bc	4.6b	7.2abc	23.7a	2.0 cd

[1] Lower case letters signify minimum significant differences within columns. Means with different letters were significantly different at $P \leq 0.05$.

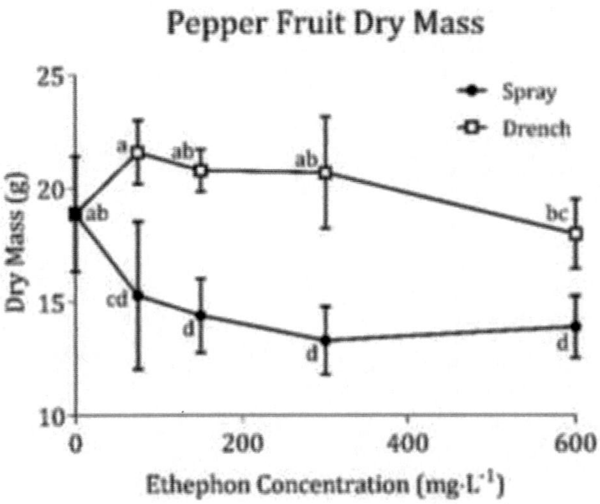

Figure 1. Pepper fruit dry mass with significant differences indicated among all treatments. Points with different lower-case letters were significantly different at $P \leq 0.05$.

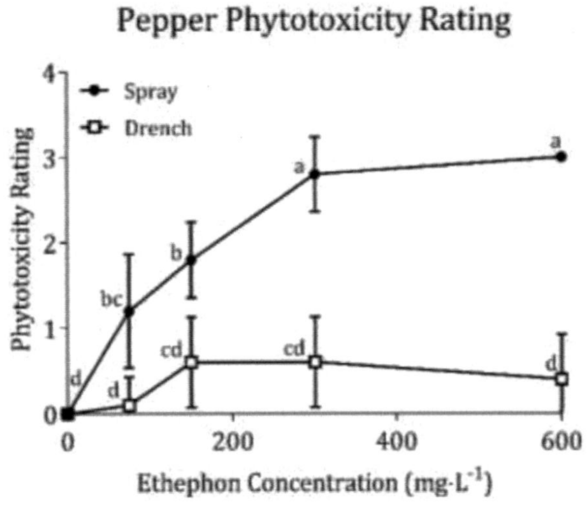

Figure 2. Pepper phytotoxicity ratings with significant differences indicated among all treatments. Points with different lower-case letters were significantly different at $P \leq 0.05$.

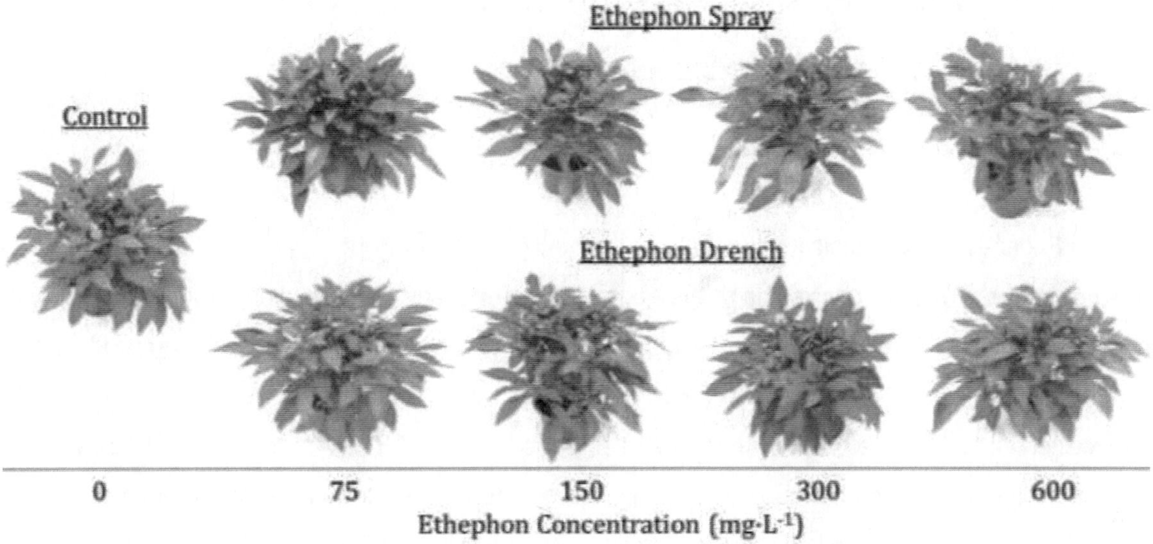

Figure 3. Representative pepper plants from each ethephon treatment compared with the untreated control plant. Phytotoxicity symptoms were observed on all sprayed plants.

SHOOTING FOR A FULLER PLANT: ANALYZING THE USE OF BENZYLADENINE AND ETHEPHON TO INCREASE RHIZOME PRODUCTION IN SANSEVIERIA

Paul Cockson, Ingram McCall, and Brian E. Whipker
Department of Horticultural Science, North Carolina State University, Raleigh, NC 27695

Foliage plants, especially indoor plants, are key pieces for interior spaces and decoration. Mother in Law's Tongue (Sansevieria sp.) is a common house plant and is grown to improve air quality and add aesthetics. Finding a method to increase in the number of rhizomes will shorten production time and create a quality, full plant for consumers. This study analyzed the effects of 6-benzyladenine (Configure; Fine Americas) and ethephon (Collate; Fine Americas) on Sansevieria laurentii to enhance rhizome production. Single shoot plants were transplanted into 6-inch pots containing a peat and perlite [80:20 (v/v)] substrate and established for three weeks. PGR drenches were applied on 24 October: 6-benzyladenine at 100, 200, 300, and 400 $mg \cdot L^{-1}$ and ethephon at 100, 200, 400, and 800 $mg \cdot L^{-1}$. After 12 weeks, the 6-benzyladenine applications did not increase the production of new shoots. At a concentration of 400 $mg \cdot L^{-1}$, the ethephon drenches resulted in an increase of shoot number by 250% as compared to the untreated control. This concentration of ethephon produced a fuller and more robust plant.

INTRODUCTION

Sansevieria is a common indoor foliage plant. Many varieties will produce chimeras which posses exotic and fascinating mutations such as variegation, striping, and unique colorations. Its natural hardiness, ornamentation, and adaptability to low light conditions makes it a perfect indoor plant. There are over 60 species of Sansevieria however, most of the commercially valuable species come from Sansevieria trifasciata (Henley et al., 1991; Griffith, 2002). Other notable species include S. cylindria and S. parva (Henley et al., 1991; Griffith, 2002). Many of the ornamental, especially variegated varieties, require propagation through cuttings, division, or rhizomes to reproduce true to type (Henley et al., 1991; Griffith, 2002).

Sansevieria naturally produces rhizomes although at a slow rate. Rhizome division has advantageous over leaf cuttings as there is a greater reservoir of nutrients in the rhizomes and roots are often present. Thus, division through rhizome is the preferred propagation method by many producers and is often the condition most material will be in when shipped to growers. Given the fact that most ornamental varieties have both a higher price and consumer demand, techniques which will enhance divisible material will be advantageous to growers. One proven way to increase plant material is by using PGRs.

In a study completed by Rayirath et al. (2009) PGRs were used to enhance rhizomes and lateral buds in rhubarb. It is a common practice to use PGRs to enhance shoots and buds when propagating plants in vitro (Banerjee et al., 2011; Roy and Banerjee, 2002). This work shows that PGRs can be utilized to enhance the production of rhizomes in some plants. Work by Miller et al. (2012) explored the use of ethephon applied as a drench rather than a floral spray. Their work showed that ethephon can be applied as a substrate drench without the threat of damage due to phytotoxicity or lack of efficiency of the chemical. The above studies show that PGR drenches can be effective and can help promote rhizome production and may have benefits when used to enhance rhizomes on sansevieria.

MATERIALS AND METHODS

Sansevieria laurentii unrooted shoots with 3 to 4 leaves were obtained (Foremostco. Inc., Miami, FL, USA) and stuck on 5 September 2016 in 1801 trays (cell size 7.93 x 7.93 x 7.62 cm deep) with an 80/20 peat/pearlite mix (Sun Gro Horticulture, Agawam, MA) and grown in a glass greenhouse at 35º N latitude in Raleigh, NC. After four weeks of rooting in a mist bench, the plants were transplanted into 6-inch diameter pots (Dillen, Middlefield, OH) and placed into a polyethylene covered greenhouse. The plants remained in the poly house for the duration of the experiment. The plants were drip irrigated with a 13-2-13 calcium + magnesium fertilizer at 150 $mg{\cdot}L^{-1}$ N.

After sufficient rooting was obtained in the 6-inch pots (3 weeks), the plants were treated on 24 Oct. 2016 with 6-benzyladenine at 100, 200, 300, and 400 $mg{\cdot}L^{-1}$ (Configure, Fine Americas) and ethephon (Collate, Fine Americas) at 100, 200, 400, and 800 $mg{\cdot}L^{-1}$ concentrations. Each plant received 4 oz (118.3 ml) applied to the substrate. The plants were randomized at six single plant replicates of each of the treatments. Data was taken ever two weeks by counting the number of above ground shoots. The experiment was terminated on 17 January 2017 and a final shoot count was obtained.

The shoot count for all treatments and reps were analyzed using SAS (version 9.4; SAS Institute, Cary, NC). The results were put through PROC GLM and the least squared means were separated using Tukey's honestly significant differences with $P \leq 0.05$.

RESULTS

After termination, the final shoot count revealed that only ethephon applied at 400 mg·L^{-1} had a statistically significant increase in rhizome proliferation (Fig. 1). The average shoot count increased by 250% when compared to the controls. At the concentrations used, the 6-benzyladenine treatments did not result in an increase in shoot production.

DISCUSSION

The results of the experiment were somewhat surprising. Given the success of PGRs on other rhizomatous plants, a better response was anticipated. One possible explanation, is that higher rates or better dosage timing was needed. Given that the higher rate of 400 mg·L^{-1} ethephon produced results, a follow up study with two species is underway which uses ethepon at higher rates and more applications.

CONCLUSION

The results of the completed experiment shows that by using ethephon at 400 mg·L^{-1}, growers can increase the amount of rhizomes by 250%. This increase in shoot proliferation will help growers immensely to produce either fuller pots for retail, or to generate a greater amount of high quality material for propagation.

ACKNOWLEDGMENT

We would like to thank Fine Americas for grant support and Sungro Horticulture for donation of the substrate used in the study.

LITERATURE CITED

Banerjee, M., Gantait, S., Pramanik, B.R. 2011. Physiology and molecular biology of plants 17(4): 387-393.

Griffith, L.P. 2002. Tropical foliage disorders. Ball Publishing, Batavia, IL.

Henley, R.W., Chase A.R., and Osborne L.S. 1991. Sansevieria production guide. Univ. of Florida, Apopka, FL.

Miller, W.B., Mattson N.S., Xie, X., Xu, D., Currey, C.J. Clemens, K.L., Lopez, R.G., Olrich, M., and Runkle, E.S. 2012. Ethephon substrate drenches inhibit stem extension of floriculture crops. HortScience 47(9): 1312-1319.

Rayirath, U.P., Rajasekaran, R.L., Caldwell, C.D., Asiedu, S.K., Sibley, K.J., Adams, A.D. 2009. CCC and prohexadione-Ca enhance rhizome growth and lateral bud production in rhubarb (Rheum rhubarbarum L.). Journal of Plant Growth Regulation 28: 137.

Roy, J., Banerjee, N. 2002. Rhizome and shoot development during in vitro propagation of Geodorum densiflorum (Lam.) Schltr. Scientia Horticulturae 94(2): 181-192.

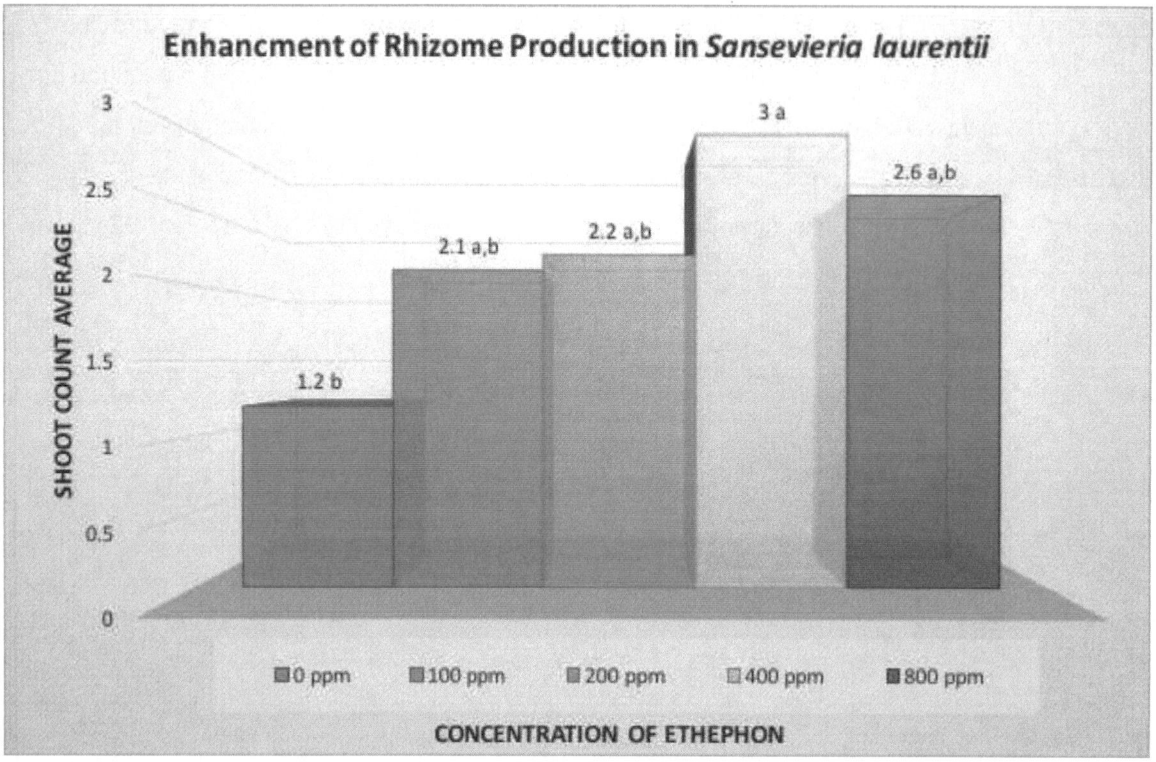

Figure 1: Effect of ethephon substrate drenches on the number of shoots produced by Sansevieria laurentii.

PERFORMANCE OF TRANSPLANTED APPLE TREES INCREASED BY IN VIVO AND IN VITRO PRODUCED ARBUSCULAR MYCORRHIZAL FUNGI

Jozsef Racsko, Stephanie Garcia

Mycorrhizal Applications, LLC. 710 NW E. Street, Grants Pass, OR 97526

Application of in vivo- and in vitro-produced mycorrhizal inocula in agricultural systems has become increasingly popular to improve crop performance. The effectiveness of inoculum in agrosystems must be assessed for the continued development of these inoculum production technologies. The goal of this study was to investigate the efficacy of in vivo- versus in vitro-produced MycoApply® inocula in container-grown apple trees. In spring of 2016, commercially available MycoApply® mycorrhizal inoculum containing four arbuscular mycorrhizal species (Rhizophagus intraradices, Funeliformis mosseae, Claroideoglomus etunicatum, Glomus aggregatum) and produced either in vivo (trade name MycoApply Endo) or in vitro (trade name MycoApply EndoMaxx) were applied to bare-root 'Honeycrisp' apple trees at a rate of 6,000 propagules per tree. Single trees were planted into 450 kg tote bags with a calcined granular clay medium and grown under nutrient and water stress. Each treatment contained 6 individual tree replicates. ANOVA analysis was used to determine statistical significance at a P level < 0.05.

Both in vivo- and in vitro-produced inocula were effective in increasing mycorrhizal colonization, vegetative growth, relative chlorophyll content, photosynthesis, and chlorophyll fluorescence. Although there were no clear effects on root volume (measured by water displacement), noticeable differences in the nutrient content were observed between the leaves of inoculated trees and the untreated trees. Mineral nutrient content in the leaves increased (N, P, Mn) and decreased (Na, Al, Cu). This study confirmed equal efficacy of in vitro- and in vivo-produced mycorrhizal inocula to improve performance of apple tree as host plant.

EFFECTS OF PLANT GROWTH REGULATORS ON GROWTH AND PHYSIOLOGICAL RESPONSES OF PERENNIAL RYEGRASS UNDER SALINITY STRESS

Zhongjie Ji, Yiwei Jiang

Department of Agronomy, Purdue University, IN 47906

Salinity stress negatively influences plant growth and physiology. Plant growth regulators (PGRs) play essential roles in regulation of plant growth and developmental processes and abiotic stress tolerance. However, salinity tolerance as affected by PGR applications are not well understood in perennial grass species. Perennial ryegrass (*Lolium perenne* L.) is a popular cool-season turf and forage grass species. The objective of this research was to investigate effects of exogenous application of PGRs on growth and physiological responses of perennial ryegrass exposed to salinity stress. Four cultivars of perennial ryegrass were exposed to 100- and 200-mM NaCl for 15d with or without PGR applications. PGRs were sprayed three times on foliage prior to salinity stress and again sprayed every other day after salinity stress was imposed. Across cultivars, plants sprayed with 6-benzylaminopunine (6-BA), Y-aminobutyric acid (GABA), and sodium nitroprusside (SNP) generally showed higher fresh weight or dry weight than the plants sprayed with water, while application of 6-BA maintained plant height under 200 mM NaCl. No changes in chlorophyll fluorescence were found between PGR-treated and untreated plants under salinity stress. The results indicated that PGRs application had positive effects on plant growth under the salinity stress. Endogenous hormone content and expression of genes involved in hormone biosynthesis are being measured to further reveal physiological and molecular mechanisms of PGRs effects on salinity tolerance in perennial grass species.

CROSS TALK BETWEEN CITRULLINE ACCUMULATION AND ABSCISIC ACID (ABA) BIOSYNTHESIS IN WATERMELONS

Vijay Joshi, Diwas Silwal, Madhumita Joshi, Kayce Noonan, Daniel Leskovar
Texas A&M AgriLife Research and Extension Center, Uvalde, TX 78801

In plants, accumulation of signaling molecules like abscisic acid (ABA) and osmolytes such as amino acids are considered to provide enhanced tolerance to abiotic stresses. Citrulline, which accumulates in high amounts in watermelon during drought stress, is regarded as a potent osmolyte and a scavenger of hydroxyl radicals. Currently little is known about the regulation of citrulline in plants. ABA has been shown to have strong positive correlations with citrulline accumulation in roots. Increasing evidences in plants suggests a role of yet uncharacterized arginine-dependent NO synthase (NOS) during drought stress which concomitantly produces citrulline along with nitric oxide (NO). As ABA is known to regulate NO synthesis, we hypothesized that citrulline accumulation in watermelon could also be regulated through arginine-dependent or independent biosynthesis. Experiments were undertaken to study the metabolic profiling and comparative expression analysis of the genes involved in the ABA and citrulline metabolism in leaves, roots and developing fruits of watermelon grown under control and drought stress treatments. Preliminary results and progress towards identifying correlations between the expressions of genes involved in the ABA-citrulline pathways and accumulation of citrulline in different plant tissues during abiotic stresses will be discussed.

CHARACTERIZATION OF THE FIRST AUXIN CONJUGATE HYDROLASE-LIKE HOMOLOGUE TO BE ISOLATED FROM LIVERWORT (MARCHANTIA POLYMORPHA) AND ITS IMPLICATIONS FOR THE EVOLUTION OF AUXIN REGULATION IN PLANTAE

James J. Campanella[1], Stephanie Kurdach[1], Joy Bochis[2], and John V. Smalley[1,2]
[1]Montclair State University, Dept. of Biology and Molecular Biology, Montclair, N.J. 07043
[2]Bergen Community College, Dept. of Biology and Horticulture, Paramus, N.J. 07652

Both vascular and non-vascular plants regulate the balance of the phytohormone auxin, primarily indole acetic acid (IAA), through interactions among de novo synthesis, degradation, efflux, influx, and conjugate synthesis/hydrolysis. IAA is stored in several conjugated forms, primarily thought not to be active. There are two major types of conjugated molecules: amide-linked IAA bound to one or more amino acids, and the ester-linked form mostly bound to sugar(s). These two types of conjugates are found at varying concentrations in the diverse tissues of Plantae (Domagalski et al., 1987). Approximately 95% of the auxin in a plant is conjugated into these storage forms (Bandurski et al., 1995; Campanella et al., 1996). Previous research has demonstrated that auxin conjugate hydrolases are pervasive in the plant kingdom to help regulate auxin activity (Campanella et al. 2003). An interspecific knowledge of the regulation and character of these pathways and their interactions is key to delineating auxin influences on plant growth and development, as well as evolution.

The main thrust of our research for the last decade has been to trace back to its earliest point the molecular evolution of this highly conserved family of hydrolase genes. This last year a high-quality version of the liverwort (Marchantia polymorpha) genome was finally available for study. We took advantage of this new liverwort data to determine if this most ancient plant species had auxin amidohydrolases. The existence of auxin regulation by means of hydrolases in the first terrestrial plants has been controversial. Ludwig-Mueller et al (2009) suggested that moss is a dead-end for the conjugate hydrolase pathway. However, Sztein et al. (1999) found that both moss and liverwort produce auxin amino-conjugates–though, there has been little evidence that hydrolases actively regulate those conjugates in either species.

Employing the new v. 3.1 liverwort genome (www.phytozome.com), we were able to detect one ILR1-like auxin conjugate hydrolase (MpILR1) in liverwort, probing with the Arabidopsis hydrolase AtILR1 gene in BLAST. The implications of this enzyme being in liverwort were quite profound because it suggested not only that there was an unbroken line of evolution for this gene family in terrestrial plants, but that auxin conjugate hydrolytic regulation may be more ancient than was thought. Also, the presence of just a single orthologue in liverwort suggests that the eventual amplification of amidohydrolase genes paralleled the rise of vascular plants.

Our main goals in this described study were to a) examine the enzymatic activity of the liverwort enzyme, b) determine its substrate recognition for auxin conjugates, c) determine its phylogenetic/evolutionary standing relative to the other "higher" plant hydrolases, and d) determine if its active site structure differs from vascular plant orthologues.

The genetic distance, neighbor-joining cladistic analysis of MpILR suggests that it is the direct ancestor of all M20 amidohydrolases found in vascular plants (Fig. 1). High bootstrap values support this conclusion. The out-group used in this phylogenetic analysis is a recently discovered auxin conjugate hydrolase from the land-sea linking algae Klebsormidium flaccidum (unpublished data). Additionally, enzymatic assays were performed employing the cloned and expressed MpILR1 protein. MpILR1 has low activity against three conjugate substrates that are hydrolyzed at levels of <10 pmol of auxin released/ min/ ml and no activity against the rest that were tested (Table 1).

Sztein et al (1999) found the most common auxin conjugates in liverwort are IAA-Valine and IAA-Glycine. The liverwort does not hydrolyze IAA-Val, but it does hydrolyze IAA-Gly at very low levels. IAA-Leu and IPA-Ala were also substrates recognized at low levels. This suggests that regulation of auxin through the hydrolysis pathway is minimal or does not occur in liverwort. Despite its low activity against amido-auxin conjugates, MpILR1 retains the basic, M20 hydrolase Mn+2 cofactor active site organization of all plant hydrolases in this family.

The MpILR1 enzyme may have low activity against auxin conjugates because it serves another physiological purpose in liverwort, or it may employ another substrate entirely. Another explanation for low activity is that the enzyme did not require greater hydrolytic function early in plant evolution—liverwort is small and slow-growing and requires little auxin to keep it growing.

To further characterize MpILR1 structure and understand activity, we analyzed its active-site for polymorphisms and found a potential change that could affect its activity against auxin conjugates. The Serine209, eventually conserved in all vascular plants, was originally a Leucine244 in the orthologous position in MpILR1 (Fig. 2). Based on the X-ray crystallographic analysis of Bitto et al. (2009), altering the Serine to a Leucine at that position results in the loss of hydrogen bonds to His380 and Asn337. Larger hydrophobic residues may not unite into the indole-binding pocket because of this structural alteration.

We have discovered an unbroken line of orthologous IAA hydrolase genes from cyanobacteria up through Charophytic algae to liverwort. We believe the early orthologue (M20 peptidase) for ILR1 has been present in the Plantae evolutionary line since cyanobacteria. We have identified orthologues in algae (K. flacciddum) and cyanobacteria (H. byssoidea). Both cyanobacteria AND the Charophytic algae appear to conserve the Leu176 residue needed for Mn+2-binding. We conclude by proposing, based on the polymorphic structure and the limited amido-auxin conjugate hydrolysis of MpILR1, that the enzyme was conserved and eventually exapted into vascular Plantae evolution.

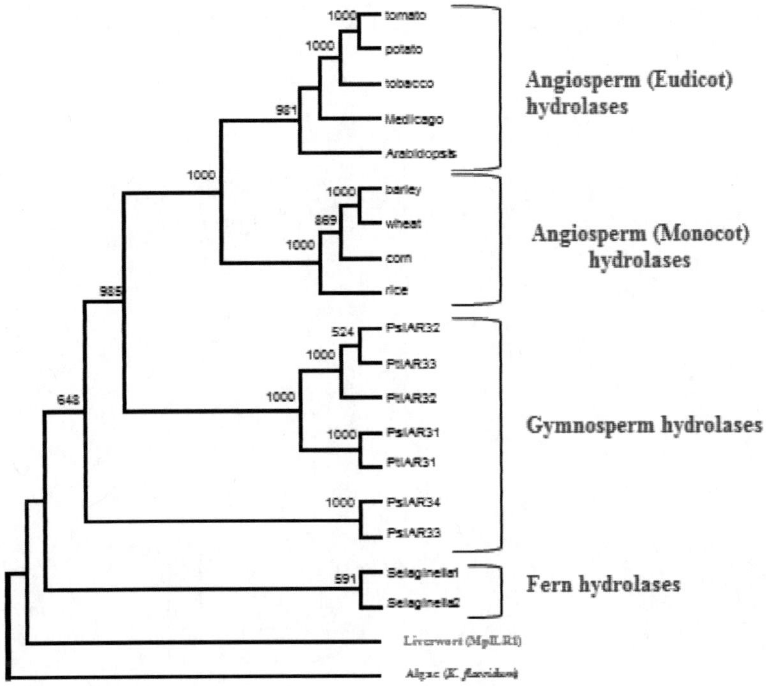

Figure 1. Neighbor-joining cladogram of the liverwort MpILR1 protein and its relationship to vascular Plantae species. The Charophytic algae K. flaccidum was employed as an out-group. 1000 bootstraps were performed.

Table 1. Enzyme/substrate specificity of the MpILR1 auxin hydrolase. All hydrolase values are in pmol auxin released/min/ml (average 4-7 replicate experiments) plus or minus the standard error. Nova Blue E. coli containing no plasmid were used as negative controls.

Enzym e	IAA-Asp	IAA-Ala	IAA-ILeu	IAA-Leu	IAA-Phe	IBA-Ala	IPA-Ala	IAA-Val	IAA-Gly
MpILR 1	0.00± 0.00	0.00± 0.00	0.00± 0.00	10.20±5. 10	0.00±0 .00	0.00±0 .00	1.66±1.6 5	0.00±0 .00	6.38±6. 25

		L244
MpILR1:	TGTVA	L
AtILR1:	FGKAA	S
		S209

Figure 2. Arabidopsis and liverwort ILR1 compared structurally in the active-site where MpILR1 shows a polymorphic change at Leucine 244 where most vascular plants hydrolases have a serine present.

PLANT BIOSTIMULANTS: AN OVERVIEW FROM THE UNITED STATES

David Beaudreau

Senior Vice President, DC Legislative and Regulatory Services, Washington, DC.

The plant biostimulant industry is growing and evolving rapidly. Many people outside of our industry do not yet have a strong understanding of what plant biostimulants are or do. As an industry, we must work together to create greater awareness of our products, to highlight their benefits and the positive roles they can play in agriculture and related industries. Biostimulants encompass a diversity of products that enhance uptake, nutrient efficiency, plant health, yield and/or quality. These naturally derived products can be used for both conventional and organic crop production.

According to U.N. estimates, as the world population continues to expand from approximately 7.5 billion today to nearly 10 billion in 2050, agriculture will need to become increasingly more efficient in order to feed a growing world. Plant biostimulants can play a role in that effort.

The global biostimulant market is estimated today to be about $1.5 billion and could grow to $4 billion by 2025. It is currently growing at an estimated 10 - 12 percent Compound Annual Growth Rate (CAGR). Both small and large companies in this space are innovative and creating new products at a considerable rate. As such, the plant biostimulant industry is a bright spot in the current agricultural economy.

ABOUT THE BIOSTIMULANT COALITION

The U.S. Biostimulant Coalition was founded in 2011 to help support the industry as it grows. Our founding statement reads, "The Biostimulant Coalition is a group of interested parties cooperating to proactively address regulatory and legislative issues involving biological or naturally-derived additives and/or similar products, including but not limited to bacterial or microbial inoculants, biochemical materials, amino acids, humic acids, fulvic acid, seaweed extract and other similar materials. The coalition is a nonprofit affiliation of interested parties." The coalition is currently comprised of 17 member companies that manufacture or distribute plant biostimulant products.

REGULATORY QUESTIONS

Our coalition was formed in part out of the need for regulatory clarity. Initial questions have been, "What are biostimulants and how can the term be defined?" The term "biostimulant" is in wide use globally, but no U.S. agency, at the state or federal level offers a working definition of biostimulants, thus no legal definition currently exists.

Many academics, industry professionals and others have attempted to draft a definition for "plant biostimulant." The European biostimulant industry and regulators have invested significant effort into drafting a definition for "plant biostimulant". The definition currently being considered: "Plant biostimulants contain substance(s) and/or micro-organisms whose function when applied to plants or the rhizosphere is to stimulate natural processes to enhance/benefit nutrient uptake, nutrient efficiency, tolerance to abiotic stress, and crop quality."

Given the lack of definitional clarity, most biostimulant products are regulated as fertilizers, soil amendments or plant inoculants at the state level. For many companies, this increases the cost and complexity of registration and compliance.

REGULATORY BODIES

There are four main regulatory bodies that either have some jurisdiction over the biostimulant industry or potentially could at some point in the future. The American Association of Plant Food Control Officials (AAPFCO), the Environmental Protection Agency, the United States Department of Agriculture and Congress all do, or could, have jurisdiction over different aspects of the plant biostimulant industry. The U.S. Biostimulant Coalition has worked to provide these entities with supporting research and other data to help clarify the existing classifications.

AAPFCO: The Association of American Plant Food Control Officials (AAPFCO) states that it seeks "to gain uniformity among each of these entities without compromising the needs of the consumers, protection of the environment or fair competition among the industry." Over the past several years, the plant biostimulant industry has spent significant time attending and participating in AAPFCO meetings.

To provide more context about AAPFCO: 1) Each state maintains its own fertilizer regulatory structure including state-specific fertilizer and soil amendment laws; 2) Definitions, standards and practices therefore differ considerably; 3) AAPFCO seeks to harmonize the diverse environment through common definitions and standards; and 4) Federal agencies, such as USDA

Natural Resources Conservation Service (NRCS), EPA and others use AAPFCO terms and definitions in their regulatory guidelines.

Some states adopt AAPFCO definitions as their standard. Other states adopt portions, but not all of the definitions, while other states follow AAPFCO direction but act independently. Despite some of the harmonization challenges, AAPFCO provides a "critical mass" for drafting terms and definitions for state-level consideration. The biostimulant industry has actively engaged with AAPFCO.

EPA: The EPA Office of Pesticide Programs has spent considerable time working on this issue since mid 2015. EPA recognizes that their determination as to what requires registration as a "pesticide" needs clarification as the science of plant health and nutrient/fertilizer efficiency has evolved. EPA is currently working to provide clarity around some of these questions, including what claims may be acceptable for biostimulants, through the creation of a guidance document.

For example, the Federal Insecticide, Fungicide and Rodenticide Act of 1972 defines pesticides to include plant regulators. The term "plant regulator" means any substance or mixture of substances, intended, through physiological action, for accelerating or retarding the rate of growth or rate of maturation, or for otherwise altering the behavior of plants or the produce thereof, but shall not include substances to the extent that they are intended as plant nutrients, trace elements, nutritional chemicals, inoculants and soil amendments.

These exclusions, and the definition of plant regulator itself, have created uncertainty for state and federal regulators, including whether certain products would fall into the excluded categories or not, how would that be determined and who would ultimately decide. Many related questions remain unanswered.

USDA: Although USDA is aware of biostimulant products, to date, it has deferred to AAPFCO for guidance on this subject. The Natural Resources Conservation Service (NRCS), a federal agency within USDA, is interested in learning more about the environmental benefits of some biostimulant products. Biostimulants could potentially be included in future USDA guidance or program documents which list best management practices focused on addressing nutrient runoff, among other topics.

CONGRESS: Given the regulatory uncertainties that exist with respect to plant biostimulants, ultimately Congress may have to pass legislation defining the term and creating a pathway toward providing regulatory clarifications. The Farm Bill or other piece of agriculturally related legislation may be a vehicle to address these issues.

GLOBAL HARMONIZATION

One of the ideal outcomes of regulatory clarifications would be global harmonization of terms and definitions surrounding the nascent and growing industry of plant biostimulants.

Europeans, in particular, see the need and are moving forward. In the European Union, significant progress has been made with respect to the classification and regulation of plant biostimulants. Beginning in 2011, European regulators and legislators began working to incorporate a broader, more holistic definition of plant nutrition/fertilizer in their fertilizer laws. They have agreed to amend existing laws to incorporate broader acknowledgement of plant biostimulants.

On July 13, 2017, the European Parliament voted to "support innovation and promote a single market for biostimulants." This recognition and acknowledgement is a positive step forward for the industry broadly and could provide a pathway and encouragement for other regulatory bodies in other parts of the world to do the same. In the long run, it will be better for regulators and the regulated community to have a simple and harmonized regulatory structure.

SUSTAINABILITY

One of the most important benefits of plant biostimulants is their ability to improve agricultural sustainability. Plant biostimulants increase the uptake and utilization of existing and applied nutrients, which in turn reduces the potential for nutrient runoff into lakes, streams and rivers. By increasing nutrient uptake, these products also reduce greenhouse gas emissions. Additionally, plant biostimulants can increase crop yield and quality without increasing the use of applied fertilizer, water or planted acres. This improves the efficient use of these natural resources and reduces food loss.

WHAT'S NEXT

We as an industry encourage the Environmental Protection Agency to issue its clarifying guidance before the end of 2017. It will be a necessary and helpful step for the industry as it continues to grow.

In the longer run, we support: 1) a definition of plant biostimulants to be included in the Farm Bill or other piece of agricultural legislation, 2) further research by USDA to demonstrated the sustainability benefits of biostimulants so growers may qualify for conservation funding and 3) a clear, consistent and predictable process for market entry.

The Biostimulant Coalition will continue to advocate for the industry and looks forward to working with all relevant stakeholders as we seek greater awareness of and better access to these agricultural tools.

CROP RESPONSES TO GIBBERELLIC ACID APPLICATION: INTERACTIONS OF CROP, SURFACTANT AND FERTILITY

Michael D. Rethwisch, Scott T. Willet, Craig J. Hruska, Kaleb Scheffler, Garrett Tooker, and Anthony Meusch

University Of Nebraska - Lincoln, Butler County Extension, David City, NE, USA 68632-1666

A number of field trials involving gibberellic acid (GA-3) application to selected area crops were conducted the past several years, with various plant growth responses noted. Soil nitrogen level was noted to limit both smooth brome brome (*Bromus inermis*) growth and resulting response to gibberellic acid application, with yield increases from gibberellic acid increasing as nitrogen level increased. Surfactants also effected crop response to gibberellic acid. Data from smooth brome natural height and extended leaf measurements indicated that best and most consistent growth responses occurred when non-ionic surfactants containing ammonium sulfate were used with gibberellic acid. These combinations resulted in positive economic hay production values across several nitrogen fertility levels in fields, especially at 60 days post treatment. Differences were also noted due to surfactant rate. In soybeans (*Glycine max*), highly significant statistical differences in stem elongation were noted among the various surfactants, including those in the same class (non-ionic surfactants that contained ammonium sulfate) by seven days post treatment. Differences in soybean growth response due to different rates of the same surfactant were not noted.

RYZUP SMARTGRASS: A NEW TOOL FOR DROUGHT PROTECTION OF CORN (ZEA MAYS)

Silverman, F. Paul[1], Falco, Kimberly[1], Wilson, Dale O. Jr.[1], Woolard, Derek[1], Surpin, Marci[1], Ott, Eric[2] and James Wargo[2]

[1]Valent BioSciences LLC, 870 Technology Way, Libertyville, IL 60048 USA

[2]Valent USA LLC, 1600 Riviera Ave, Walnut Creek, CA 94596 USA

RyzUp SmartGrass® (GA3, 40% SG formulation) is a foliar treatment that increases plant growth, grain, and silage yields in the United States. It is used extensively on corn, wheat and pasture grasses and offers increases in plant vigor, growth rates, and grain yield. In 2012, the Midwestern U.S. experienced unseasonable, severe drought. This drought negatively affected grain and silage yields. Application of RyzUp Smartgrass at V2-V6 timing at rates of 3.0-6.2 g of GA3 in 15-25 gallons spray volume per acre effectively mitigated a percentage of the yield losses experienced by corn. In multiple trials, application of RyzUp Smartgrass caused a dose-dependent increase in hybrid corn yield under drought. Moreover, application of GA3 to hybrid corn also increased silage quality. Mitigation of drought by RyzUp Smartgrass is likely due to multiple mechanisms, which will be discussed.

APPLICATION OF FRESCO FOLIAR SPRAYS TO COUNTERACT PACLOBUTRAZOL RESIDUE IN IRRIGATION WATER: AND THE IMPLICATIONS OF UTILIZING MICRO-DRENCHES TO CONTROL PLANT GROWTH

Brian E. Whipker, and Ingram McCall

Department of Horticultural Science, North Carolina State University, Raleigh, NC 27695 U.S.A.

Paclobutrazol (paclo) is the most widely applied plant growth regulator (PGR) used in commercial greenhouses in the U.S. In larger operations, paclo can be applied via an overhead boom as a drench. This results in a significant amount of the PGR ending up on the floor and consequently in the irrigation water that is collected for reuse. This low level of residual paclo can result in excessive control of sensitive ornamental species. One potential option to counteract the effects of the residual paclo is to apply Fresco (a combination of GA4+7+6 BA). Wax begonias were grown with 0, 5, 10 or 50 ppb paclo, and then treated twice with Fresco foliar sprays of 0, 1, 3 or 6 ppm. The use of Fresco stimulated growth to counteract the anti-GA effects of the paclo. Effects were greatest at the lowest levels of paclo. Results offer an alternative to growers for overcoming low dose paclo residue in the irrigation water. In addition, plants grown with 0, 5, 10 or 50 ppb paclo, but not treated with Fresco, illustrate the potential of how low dose micro-drenches can be utilized by greenhouse operators as a growth control option.

INTRODUCTION

Plant growth regulators (PGRs) are effective tools for controlling excessive plant growth. PGRs can be applied as foliar sprays, substrate drenches, a high volume foliar spray commonly referred to as a sprench, a preplant liner root dip, or a preplant bulb soak (Latimer and Whipker, 2012). All these methods are employed by greenhouse operations in order to produced well-proportioned plants to the size of the pot.

Paclobutrazol is one of the most widely used PGRs on ornamental plants (Whipker, 2017). Applications of PGRs come in contact with the targeted plants or pot, but with the applications used, the excess can end up being applied to the concrete floor. For larger greenhouse operations that collect and recirculate the irrigation water, this excess material can be washed into the water reservoir and reapplied to the plants (Adriansen and Odgaard, 1997). Levels of over 50 parts per billion have been reported in irrigation water (Grant et al., 2016). For sensitive species such as begonias, pansies, or vinca, overall plant growth can be severely stunted.

Removal techniques have been evaluated by Grant et al. (2016) in which they reported that activated carbon to be an effective product to remove ~90% of the residual palcobutrazol. PGRs which contain gibberellins A4+A7 and N-(phenylmethyl)-1H-purine-b-amine such as Fresco (Fine Americas) or Fascination (Valent USA) are used by greenhouses operations to enhance plant growth (Whipker, 2017). This research investigated the use of Fresco foliar sprays at overcoming the growth control effects of paclobutrazol applied at low concentrations in the irrigation water.

MATERIALS AND METHODS

'Super Olympia Rose' begonia plugs (144 size; 2.6 x 2.6 x 4 cm) were transplanted into 1801 cell packs containing Fafard 1P substrate on 10 October 2014. The plants were grown in a glass greenhouse at 35°N latitude in Raleigh, NC. The temperatures were 18 °C nights and 24 °C days. The plants were randomly assigned one of four groups. All plants were fertilized at every irrigation with 17N–1.7P–14.1K mixed at 100 mg·L^{-1} nitrogen. In addition, each of the four groups had either 0, 5, 10 or 50 ppb of paclobutrazol (Piccolo 10XC, Fine Americas) mixed into the fertilizer solution. This was to simulate the presence of paclobutrazol in irrigation water at representative concentrations reported by Grant et al. (2016). Plants were overhead watered with the fertilizer solution 10 times over the course of the experiment.

To counteract the effects of the paclobutrazol in the irrigation water, four Fresco foliar spray treatments was applied to each of the paclo treated groups. Concentrations of 0, 1, 3, or 6 mg·L^{-1} gibberellins A4+A7 and N-(phenylmethyl)-1H-purine-b-amine (Fresco; Fine Americas). Fresco was applied on 23 and 30 October. Sprays were applied to the foliage and fruit at a rate of 203.7 mL·m^{-2}, and were sprayed to the point of slight runoff.

The experiment was completely randomized with six single plant replicates of each of the 16 treatments. The experiment was terminated on 10 November. Plants were measured for height and two diameters. Data for plant height and diameter were analyzed using SAS (version 9.4; SAS Institute, Cary, NC). Data were subjected to PROC GLM and the least squared means were separated by Tukey's honestly significant differences at $P \leq 0.05$.

RESULTS AND DISCUSSION

Plant growth was smaller as the constant concentration of paclobutrazol increased from 0 to 50 ppb (Fig. 1). The low concentration of paclobutrazol caused a significant stunting of the begonia plants at 50 ppb. Wax begonias are very sensitive to triazole PGRs (Whipker, 2017). The foliar applications of Fresco counteracted the anti-GA effects of the low dose paclobutrazol. As the concentration of Fresco increased, the plants were larger (Fig. 2&3, Table 1). The effects were greatest at the highest concentrations of Fresco and the lowest concentrations of paclobutrzol. Growth stimulation was the least when the plants received 50 ppb paclobutrazol. Growers will need to determine the amount of residue paclobutrazol that they have in their recirculating irrigation water and then select a suitable Fresco application concentration. The double application of 6 ppm Fresco resulted in lighter colored new growth that was slightly stretched (Fig. 4). This rate would be excessive if lower concentrations of paclobutrazol were present in the irrigation water.

CONCLUSIONS

Fresco foliar sprays were effective in promoting new growth and overcoming the growth suppression caused by the low ppb dose of paclobutrazol in the irrigation water. Growers can use Fresco foliar sprays as a method to overcome a paclobutrozol residue problem and it offers another option instead of investing into an activated charcoal filtration system.

Alternatively, the amount of growth control that occurs with low dose ppb continual drenches may offer an application option for greenhouse producers. Prior work by Barrett (2008) with 0.1 ppm drenches of paclobutrazol, applied 3 to 6 times during the production cycle of poinsettias, provided ample control of plant growth. A dose of 0.1 ppm is 100 ppb. Therefore, for plants that are less sensitive to PGR applications, a 50 to 100 ppb (0.05 to 0.1 ppm) continuous application of paclobutrazol may offer a control option to greenhouse growers.

LITERATURE CITED

Adriansen, E. and P. Odgaard. 1997. Residues of paclobutrazol and uniconazole in nutrient solutions from ebb and flood irrigation of pot plants. Scientia Horticulturae 69:73-83.

Barrett, J. 2008. Control poinsettia size with early drench applications. GPN. http://www.gpnmag.com/article/control-poinsettia-size-early-drench-applications/

Grant, G.A., P.R. Fisher, J.E. Barrett, and C.P. Wilson. 2016. Remediating paclobutrazol from irrigation water using activated carbon. https://ashs.confex.com/ashs/2016/webprogram/Paper24570.html

Latimer, J. and B.E. Whipker. 2012. Selecting and using plant growth regulators on floricultural crops. Virginia Tech Coop. Ext. Serv. Publication 430-102, pp. 32.

Whipker, B.E. Plant growth regulators for annuals. Ball Publishing, West Chicago, IL, pp. 52.

Table 1. Data for the growth response of growth stimulation which resulted from the foliar application of Fresco at 0, 1, 3 or 6 ppm applied twice to counteract the two low dose concentrations of paclobutrazol (0, 5, 10, or 50 ppb) supplied in the irrigation water.

Begonia
'Super Olympia Rose'

Begonia Plant Height

	0 Fresco	1 ppm Fresco	3 ppm Fresco	6 ppm Fresco
0 Paclo	6.5	9.2	7.4	8.0
5 ppb Paclo	5.5	8.1	8.1	9.0
10 ppb Paclo	5.3	6.9	7.3	9.9
50 ppb Paclo	4.4	6.0	6.5	7.1

Plant height (cm) significant by the interaction of Paclo Irrigation treatment x Fresco foliar spray rate, at 0.0001.

Figure 1. Response of wax begonias to low dose concentrations of paclobutrazol (0, 5, 10, or 50 ppb) added to the irrigation water and applied at each irrigation.

Figure 2. Side view of the growth stimulation which resulted from the foliar application of Fresco at 0, 1, 3 or 6 ppm applied twice to counteract the two low dose concentrations of paclobutrazol (0, 5, 10, or 50 ppb) supplied in the irrigation water.

Figure 3. Top view of the growth stimulation which resulted from the foliar application of Fresco at 0, 1, 3 or 6 ppm applied twice to counteract the two low dose concentrations of paclobutrazol (0, 5, 10, or 50 ppb) supplied in the irrigation water.

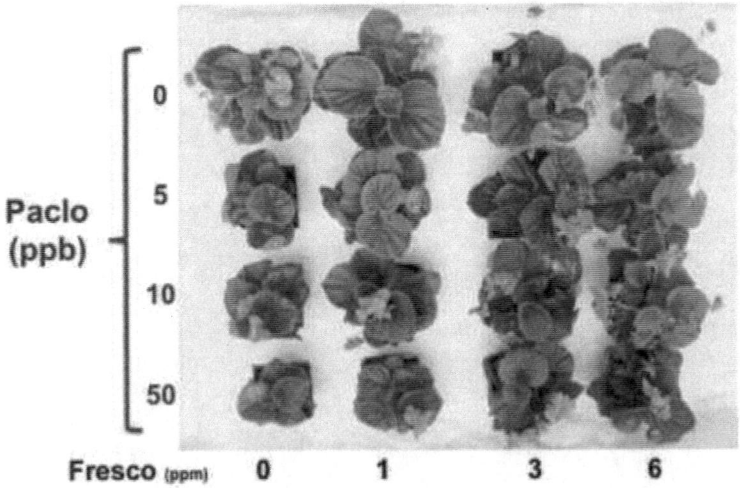

The application of Fresco foliar sprays (ppm) counteracted the low level of paclobutrazol (ppb) in the irrigation water.

Figure 4. Excessive growth that resulted from two applications of Fresco foliar sprays at 6 ppm.

EFFECTS OF PLANT GROWTH REGULATORS ON SEED GERMINATION AND ROOT GROWTH TRAITS OF RED AND YELLOW ONION

Maria Andrea Macias[1,2], Daniel I. Leskovar[2]

[1]Texas A&M Horticulture Department (mandreaml@tamu.edu)

[2]Texas A&M AgriLife Research (Texas A&M University) Uvalde, 78801 Texas, USA

E-mail: d-leskovar@tamu.edu

Onions (Allium cepa) are easily outcompeted by weeds due to slow germination and relative growth rate, which results in bulb size reduction and less homogenous populations. Therefore, high seed germination and root vigor are important traits to improve stand establishment and field performance of onion. The effects of exogenous plant growth regulators (PGRs), 2-chloroethylphosphonic acid (ethephon, Eth), 3-idoleacetic acid (Auxin, IAA), trans-zeatin (CTK) and 1-aminocyclopropane-1-carboxylic acid (ACC) were evaluated on germination and early root growth of onion seedlings. Seeds were soaked for ten hours in 10, 30, 100μM ethephon and ACC; 1, 3, 10μM IAA and CTK; 100μM ethephon + 10μM IAA (Ethephon+IAA), combination; and water (hydro priming). Seed germination was significantly improved by the application of 10 to 100μM Eth, Ehtephon+IAA, and 3μM IAA. Root surface area was increased by Eth (range of 10-100μM), Ethephon+IAA and 3μM IAA. Root length and average diameter was enhanced by 1μM CTK and 100μM ACC. Conversely, Eth reduced primary root growth and average root diameter, while IAA showed no effect on these root traits. A subsequent experiment evaluated the synergistically effects of combinations of hormones. Seeds treated with ACC (250 μM) + CTK (0.5μM) and ACC (250 μM) + CTK (0.5μM) + Eth (20 μM), displayed primary root length and average root diameter increased in both cultivars. Root surface area was not affect for any of the hormone combinations. Our results suggest that exogenous PGRs could be useful to enhance germination, root length and surface area of onion seedlings, which in turn may improve early growth after stand establishment.

SUSTAINABILY ON LONG DURATION SPACE MISSIONS: PLANTS, MICROBES AND A CONTROLLED ENVIRONMENT

Gary W. Stutte, Ph.D
SyNRGE, LLC, Exploration Park, Florida 32953

The development of sustainable plant growth systems will be required for long duration space missions in order to maintain air quality, purify water and produce food. Sustainability will be achieved through crop selection, environmental control and rhizosphere management. For example, CO_2 increases crop yield, reduces harvest time, and may improve quality. Light emitting diodes (LEDs) enable spectral quality to be used to manage plant morphology and increase the production of bioactivity phytochemicals that improve quality. Microgravity and spacecraft environment impose unique stresses that increase risk of abiotic and biotic stresses reducing yield. The selection beneficial, symbiotic and mutualistic microbial organisms suggest that resistance to stress is increased, yields improved, and increased system stability achieved. Understanding the interactions crops, microbial interactions, and environmental conditions is necessary of future space missions. It also has direct application to improving yield and quality of crops in controlled environments, including greenhouses, plant factories and vertical farms.